Power Systems

Electrical power has been the technological foundation of industrial societies for many years. Although the systems designed to provide and apply electrical energy have reached a high degree of maturity, unforeseen problems are constantly encountered, necessitating the design of more efficient and reliable systems based on novel technologies. The book series Power Systems is aimed at providing detailed, accurate and sound technical information about these new developments in electrical power engineering. It includes topics on power generation, storage and transmission as well as electrical machines. The monographs and advanced textbooks in this series address researchers, lecturers, industrial engineers and senior students in electrical engineering.

Power Systems is indexed in Scopus

Wesley Pacheco Calixto •
Wanderson Rainer Hilário Araújo •
Lucas Diniz Silva Morais •
Marcio Rodrigues Cunha Reis

Control and Tracking Techniques for Switched Reluctance Machines

Springer

Wesley Pacheco Calixto
Electrical, Mechanical & Computer Engineering School (Federal University of Goias); Technology Research and Development Center (GCITE)
Federal Institute of Goias
Inhumas, Goiás, Brazil

Lucas Diniz Silva Morais
Technology Research and Development Center (GCITE)
Federal Institute of Goias
Senador Canedo, Goiás, Brazil

Wanderson Rainer Hilário Araújo
Graduate Program in Production and Systems Engineering
Pontifical Catholic University of Goias
Goiânia, Goiás, Brazil

Marcio Rodrigues Cunha Reis
Technology Research and Development Center (GCITE)
Federal Institute of Goias
Senador Canedo, Goiás, Brazil

ISSN 1612-1287 ISSN 1860-4676 (electronic)
Power Systems
ISBN 978-3-031-86726-2 ISBN 978-3-031-86727-9 (eBook)
https://doi.org/10.1007/978-3-031-86727-9

This Springer imprint is published by the registered company Springer Nature Switzerland AG
The registered company address is: Gewerbestrasse 11, 6330 Cham, Switzerland

We dedicate this work to the remarkable professors who, with their vision and enthusiasm, inspired us, believed in our potential, and envisioned our future achievements. This work is also for those who supported us in conducting research, developing teaching projects, and providing guidance that made our academic achievements possible. We recognize that such mentors and collaborators are rare, but their influence is immeasurable. To them, we extend our eternal gratitude for being an integral part of our journey.

Preface

Imagine a new landscape in electrical engineering,[1,2] where innovation reshapes how we manage energy and control systems. In this ever-evolving field, switched reluctance machines (SRMs) emerge as key players, offering resilience, efficiency, and versatility across applications ranging from speed control to sustainable energy generation. This book invites you to delve into the world of SRMs through an integrated and accessible journey, bridging theoretical foundations, technological advancements, and practical implementations.

Through meticulously structured chapters, you will uncover the core of this technology, learning how recent developments have positioned SRMs as competitive solutions in both industrial and energy sectors. With unique features such as real-world case studies and experimental model validation, this work stands as an indispensable reference for engineers, students, and industry professionals. It equips you with the tools and knowledge necessary to master this technology and contribute to a future of greater efficiency and sustainability.

Accept this invitation and be inspired by the possibilities SRMs can unlock. Page by page, discover how this technology is shaping the future of electrical engineering, empowering you to turn challenges into opportunities and innovation into progress.

Inhumas, Brazil — Wesley Pacheco Calixto
Goiânia, Brazil — Wanderson Rainer Hilário Araújo
Senador Canedo, Brazil — Lucas Diniz Silva Morais
Senador Canedo, Brazil — Marcio Rodrigues Cunha Reis

[1] Prof. Dr. Cleber Asmar Ganzaroli, Goiano Federal Institute.

[2] Prof. Dr. Luiz Eduardo Bento Ribeiro, Federal Institute of Goias.

Acknowledgments

We extend our deepest gratitude to all those who, through generosity, expertise, and dedication, have enriched this work and made it more complete. Each contribution has been an indispensable element in the creation of this book, reflecting the collective efforts of brilliant and committed minds. We thank Prof. Dr. Viviane Margarida Gomes Pacheco for her invaluable contribution in providing images and collaborating on the writing of sections of the text, her participation serving as a true source of inspiration. Professor M.Sc. Flávio Adalberto Gomes, whose meticulous review of the preliminary sections and Chap. 1 enhanced the quality of this work. The authors thank Prof. Dr. Geovanne Pereira Furriel for his thorough review of Chap. 3, demonstrating exceptional attention to detail and a commitment to academic excellence. The authors thank Prof. Dr. Saulo Rodrigues e Silva for his critical review of Chap. 4, reflecting intellectual rigor and academic generosity. We thank Prof. Dr. Luiz Eduardo Bento Ribeiro for authoring the Preface and reviewing Chap. 5, whose insightful contributions and introductory words have significantly enriched this book. The authors thank Prof. Dr. Cleber Asmar Ganzaroli for authoring the Preface and reviewing Chap. 6, whose efforts added both value and impact to the text. We express our sincere gratitude and admiration for your invaluable contributions. This book embodies the spirit of collective effort, shared wisdom, and collaboration that we deeply appreciate and honor. It would not have been the same without your support.

Competing Interests The authors have no competing interests to declare that are relevant to the content of this manuscript.

Organization of the Book

The development of this work stems from years of dedication to research, learning, practical applications and projects conducted in classrooms and laboratories. This continuous effort enabled the design and construction of the switched reluctance machine (SRM), consolidating accumulated knowledge, and bringing this work to fruition. From the outset, significant challenges were encountered, such as during the construction of the experimental platform for machine control and operation, where the generator's voltage control required additional components, complicating adjustments and hindering the alignment of results with the computational model.

A key highlight of this book is the development of the inductance surface using parametric regression, which allowed for simulations that closely align with real-world behavior while reducing computational effort. Another notable achievement is the determination of the instantaneous torque value during machine operation, a critical aspect for understanding the generator's efficiency, which depends on the torque provided by the primary machine. Additionally, sensitivity analysis played a pivotal role in defining the variable to be manipulated by the Proportional–Integral–Derivative (PID) controller. This controller, designed to handle reference variations, load disturbances, and changes in switching angles, required non-conventional tuning achieved through optimization methods.

This work provides a comprehensive and progressive approach to switched reluctance machines (SRMs), combining theoretical foundations with practical perspectives in eight chapters organized to explore their origins, principles, models, and applications. Chapter 1 introduces the evolution of SRMs, Chap. 2 discusses their technical characteristics and control strategies, and Chap. 3 delves into theoretical and computational models with an emphasis on optimization. Chapter 4 focuses on parametric regression and sensitivity analysis, while Chap. 5 details the construction of experimental platforms and validation procedures.

Chapter 6 presents approaches for optimizing generator models, Chap. 7 examines the performance of SRMs as motors, and Chap. 8 highlights their applications in renewable energy, with a particular focus on wind energy systems. Suitable for researchers, engineers, and students, this book offers a deep and practical understanding of SRMs, establishing itself as a valuable reference in contemporary engineering.

Contents

List of Abbreviations

AC	Alternating Current
AE	Absolute Error
AHB	Asymmetric Half Bridge
AI	Artificial Intelligence
BIBO	Bounded Input and Bounded Output
DC	Direct Current
GA	Genetic Algorithm
HAWT	Horizontal Axis Wind Turbine
IAE	Integral of the Absolute Error
IAS	Industry Applications Society
IEEE	Institute of Electrical and Electronics Engineers
IGBT	Insulated Gate Bipolar Transistor
IM	Induction motor
LWT	Large Wind Turbine
MEPT	Maximum Efficiency Point Tracking (Turn-Off Angle Tracking)
MPPT	Maximum Power Point Tracking
QNM	Quasi Newton Method
GA-QNM	Genetic Algorithm Hybridized with Quasi Newton Method
PI	Proportional-Integral
PID	Proportional-Integral-Derivative
PMSG	Permanent Magnet Synchronous Generator
PWM	Pulse Width Modulation
RMS	Root Mean Square
SRG	Switched Reluctance Generator
SRGA	Self-Excited Switched Reluctance Generator
SRGIE	Independently Excited Switched Reluctance Generator
SRM	Switched Reluctance Machine

List of Figures

List of Tables

Chapter 1
Brief Historical Overview of Switched Reluctance Machines

1.1 Historical Development and Evolution

The technology of switched reluctance machines (SRMs) was developed by William Hannis Taylor in 1838 in the United States and patented in 1840 in England. Based on the principle of magnetic reluctance, SRMs utilize rotor movement to minimize circuit reluctance in response to the excitation of stator windings [1, 2]. The first SRM was constructed in Scotland for use in locomotives, enabling the conversion of electrical energy into mechanical motion [3, 4]. During the first half of the 19th century, this technology was studied to improve electric traction systems. However, the lack of electronic switching technologies limited its initial development. In the second half of the 20th century, advancements in power electronics and microprocessor technology facilitated the rediscovery and refinement of SRMs, establishing them as a viable and competitive alternative to conventional electric motors [5, 6].

Research conducted in the late 20th century, particularly at the universities of Nottingham and Leeds, was crucial for advancing SRM technology by establishing the theoretical foundations for its control and promoting its use in traction and industrial drive applications. The combination of operational resilience, efficiency, and the ability to operate at high speeds in adverse conditions enabled SRMs to excel in sectors such as automotive, aerospace, and distributed generation [7, 8].

Despite challenges such as acoustic noise and torque ripple, continuous improvements in control algorithms and the development of innovative topologies, including segmented SRMs and hybrid models with permanent magnets, have led to significant performance advancements [9, 10]. In addition, the adoption of multilevel converters and the implementation of advanced control strategies have further mitigated these issues, positioning SRMs as a promising solution for electric vehicles, aircraft systems, and other applications that require high power density and reliability [11, 12].

W. P. Calixto et al., *Control and Tracking Techniques for Switched Reluctance Machines*, Power Systems, https://doi.org/10.1007/978-3-031-86727-9_1

In recent decades, SRMs have evolved from an experimental technology to a consolidated solution for various industrial and energy applications. Their efficiency, low maintenance, and adaptability to diverse operating conditions make them a competitive alternative to conventional motors [13, 14]. Current research focuses on overcoming the remaining challenges related to large-scale production and performance optimization, further establishing SRMs as a strategic technology for the future of electrification and sustainability across multiple sectors [15].

1.2 Switched Reluctance Motors: Advantages and Advancements

SRMs, when used as motors, offer simplicity in construction, low manufacturing cost, and efficient performance at a wide range of speeds [9]. Their operation relies on sequential stator-coil energization, creating magnetic flux that guides the rotor to positions of minimal reluctance. This operation results in high torque density and excellent fault tolerance, making SRM motors ideal for applications that require operation in harsh conditions [16].

Recent advances in control techniques and innovative topologies, such as segmented SRM and hybrid configurations, have improved the dynamic performance and efficiency of these machines [17]. Although conventional SRM motors lack permanent magnets, hybrid variations have been developed to enhance torque density and overall efficiency through the incorporation of these magnetic components. These hybrid topologies complement the advantages of conventional SRM motors while addressing issues such as torque ripple and acoustic noise [10, 11].

The simple construction of conventional SRM motors, which lack permanent magnets and rotor windings, results in lower production and maintenance costs [9]. This characteristic, combined with high reliability and operational flexibility, makes SRM motors an attractive option for industrial automation, electric and hybrid vehicles, and other applications that demand energy efficiency and low maintenance costs. In the automotive sector, the ability of SRM motors to operate as both a motor and a generator facilitates the seamless transition between traction and energy generation, resulting in significant efficiency gains [18, 19].

With increasing adoption in the automotive, aerospace, and distributed generation sectors, SRM motors continue to evolve to meet the demand for more efficient and sustainable solutions. Current research focuses on optimizing control and the exploration of new topologies, further establishing SRM motors as a strategic technology for the future of electrification in various industrial and energy sectors [10, 20].

1.3 Switched Reluctance Generators: Efficiency and Control

When applied as generators (SRGs), SRMs have become a viable option for variable-speed applications, especially in wind energy systems, due to their simplicity, efficiency, and reliability. The SRG converts mechanical energy into electrical energy through variations in magnetic reluctance within the magnetic circuit, allowing efficient operation over a wide range of speeds and under varying load conditions [21, 22].

The SRG's efficiency is highly dependent on precise switching angle control and excitation voltage regulation. Advanced control strategies, such as current regulation and torque control, have been introduced to mitigate challenges such as torque ripple and acoustic noise [23]. In addition, maximum power point tracking (MPPT) algorithms are widely used to optimize energy extraction under variable operating conditions, such as changes in wind speed and load variations [9, 24].

The simple construction and the absence of rare materials make the SRG a competitive alternative compared to conventional generators, such as permanent magnet synchronous generators (PMSG) and doubly-fed induction generators (DFIG), which are more complex and costly to manufacture and maintain. Current research has focused on new configurations, such as designs with a higher number of rotor poles relative to stator poles and magnetically independent stator–rotor sets, to improve torque characteristics and operational flexibility [25].

Despite these advantages, the use of SRGs is still less widespread than that of other commercial technologies. Recent studies have focused on SRG modeling and simulation to evaluate its electromagnetic behavior and propose new methods for voltage and power control. Precise control of the position of the rotor and optimization of drive parameters are critical areas for improving the dynamic performance and energy efficiency of SRGs [21, 26].

With advances in control techniques and ongoing development in design and optimization, the SRG continues to establish itself as a strategic technology to meet the demand for sustainable energy generation solutions [27]. The development of new control methods and improvements in the SRG topology aim to make it even more efficient and reliable for applications in distributed generation and renewable energy sources, such as wind farms. In this context, the SRG positions itself as a competitive and sustainable alternative for the future of clean and efficient energy generation [28, 29].

1.4 Technological Innovations and Emerging Applications

Switched reluctance machines (SRMs) have established themselves as an emerging technology capable of meeting the growing demand for renewable and sustainable energy sources [10]. Recent advances have allowed SRMs to be explored in new applications, such as energy storage systems and high-precision industrial drives.

Their efficient performance under varying speeds and load conditions makes them suitable for kinetic energy storage and robotic motion control, where high dynamic performance and quick response are critical [13].

Advancements in structural design, such as the adoption of new high-permeability magnetic materials and optimized topologies, have enhanced the efficiency and power density of SRMs, while reducing core losses and enabling better thermal management. These improvements have expanded the use of SRMs in precision drives for medical equipment and in the automation of critical industrial processes, where stability and precise torque control are paramount [9, 30].

The introduction of adaptive control algorithms using artificial intelligence and neural network-based optimization has improved the response of SRMs to load variations and shifting operational conditions [9, 10]. Emerging applications include the use of SRMs in drones and autonomous vehicles, leveraging their ability to operate efficiently under high-speed and variable-load scenarios. Machine learning algorithms are being applied to predict failures and optimize real-time operation, increasing the lifespan and reliability of machines [31].

In the field of renewable energy generation, switched reluctance generators (SRGs) have been used to maximize energy extraction under low wind speeds, variable climatic conditions, and for integration with hybrid systems, such as solar panels, which form intelligent microgrids [32]. The versatility of SRGs and their ability to operate under harsh environmental conditions make them a promising option for remote regions or areas with limited infrastructure, where distributed generation is critical for local energy development [21, 26].

Beyond conventional applications, recent research has investigated the use of SRMs in propulsion systems for electric aircraft, exploring their high power density and fault tolerance. Integration with new storage technologies, such as supercapacitors and solid-state batteries, is also being studied to optimize energy management in electric vehicles, enabling faster charging cycles and extended range [19, 33].

These advancements position SRMs as a promising and competitive alternative in a growing number of emerging applications. Future research should focus on the development of hybrid topologies and the integration of autonomous control technologies, further establishing SRMs as a strategic solution for the future of electrification, sustainable mobility, and distributed energy generation [2, 34].

References

1. S.S. Ahmad, G. Narayanan, Evaluation of DC-link capacitor RMS current in switched reluctance motor drive. IEEE Trans. Ind. Appl. **57**(2), 1459–1471 (2020)
2. L. Sufei et al., Modeling, design optimization, and applications of switched reluctance machines: a review. IEEE Trans. Ind. Appl. (2019). https://doi.org/10.1109/TIA.2019.2897965
3. K. Vijayakumar et al., Switched reluctance motor modeling, design, simulation, and analysis: a comprehensive review. IEEE Trans. Magn. (2008). https://doi.org/10.1109/TMAG.2008.2003334

4. C. Kamalakannan et al., Switched reluctance machine in automotive applications —a technology status review. in *2011 1st International Conference on Electrical Energy Systems* (2011). https://doi.org/10.1109/ICEES.2011.5725326
5. B. Bilgin et al., Making the case for switched reluctance motors for propulsion applications. IEEE Trans. Veh. Technol. (2020). https://doi.org/10.1109/TVT.2020.2993725
6. N. Zabihi R. Gouws, A review on switched reluctance machines for electric vehicles. in *International Symposium on Industrial Electronics* (2016). https://doi.org/10.1109/ISIE.2016.7744992
7. W.F. Ray et al., High performance switched reluctance brushless drives, in *1985 Annual Meeting Industry Applications Society* (1985)
8. W.F. Ray et al., High-performance switched reluctance brushless drives. IEEE Trans. Ind. Appl. **IA-22**(4), 722–730 (1986) https://doi.org/10.1109/tia.1986.4504784
9. G. Fang et al., Advanced control of switched reluctance motors (SRMs): a review on current regulation, torque control and vibration suppression. IEEE Open J. Ind. Electron. Soc. **2**, 280–301 (2021) https://doi.org/10.1109/ojies.2021.3076807
10. S. Li et al., Modeling, design optimization, and applications of switched reluctance machines: a review. IEEE Trans. Ind. Appl. **55**(3), 2660–2681 (2019) https://doi.org/10.1109/tia.2019.2897965
11. Z. Xu et al., A review on segmented switched reluctance motors. Energies **15**(23), 9212 (2022). https://doi.org/10.3390/en15239212
12. C. Donaghy-Spargo, Synchronous reluctance motor technology: industrial opportunities, challenges and future direction. Eng. Technol. Ref. (2016). https://doi.org/10.1049/etr.2015.0044
13. K. Diao et al., Design optimization of switched reluctance machines for performance and reliability enhancements: a review. Renew. Sustain. Energy Rev. **168**, 112785 (2022). https://doi.org/10.1016/j.rser.2022.112785
14. M. Abdalmagid et al., Geometry and topology optimization of switched reluctance machines: a review. IEEE Access **10**, 5141–5170 (2022). https://doi.org/10.1109/access.2022.3140440
15. M. Murataliyev et al., Synchronous reluctance machines: a comprehensive review and technology comparison. Proc. IEEE **110**(3), 382–399 (2022). https://doi.org/10.1109/jproc.2022.3145662
16. M.M. Bouiabady, A.D. Aliabad, E. Amiri, Switched reluctance motor topologies: a comprehensive review, in *Switched Reluctance Motor: Concept, Control and Applications*. (InTech, 2017). ISBN: 9789535132677. https://doi.org/10.5772/intechopen.69149
17. W.R.H. Araujo et al., Efficiency enhancement of switched reluctance generator employing optimized control associated with tracking technique. Energies **14**(24), 8388 (2021). https://doi.org/10.3390/en14248388
18. J.D. Widmer, R. Martin, B.C. Mecrow, Optimization of an 80-kW segmental rotor switched reluctance machine for automotive traction. IEEE Trans. Ind. Appl. **51**(4), 2990–2999 (2015). https://doi.org/10.1109/tia.2015.2405051
19. B. Bilgin et al., Making the case for switched reluctance motors for propulsion applications. IEEE Trans. Veh. Technol. **69**(7), 7172–7186 (2020). https://doi.org/10.1109/tvt.2020.2993725
20. M. Krishnamurthy et al., Making the case for applications of switched reluctance motor technology in automotive products. IEEE Trans. Power Electron. **21**(3), 659–675 (2006). https://doi.org/10.1109/tpel.2006.872371
21. F.P. Scalcon et al., A Review on switched reluctance generators in wind power applications: fundamentals, control and future trends. IEEE Access **10**, 69412–69427 (2022). https://doi.org/10.1109/access.2022.3187048
22. R. Cardenas et al., Control of a switched reluctance generator for variable-speed wind energy applications. IEEE Trans. Energy Conversion **20**(4), 781–791 (2005). https://doi.org/10.1109/tec.2005.853733
23. A. Arifin, I. Al-Bahadly, Switched reluctance generator for variable speed wind energy applications. Smart Grid Renew. Energy **2**(1), 27–36 (2011). https://doi.org/10.4236/sgre.2011.21004

24. T.A. dos Santos Barros et al., An approach for switched reluctance generator in a wind generation system with a wide range of operation speed. IEEE Trans. Power Electron. **32**(11), 8277–8292 (2017). https://doi.org/10.1109/tpel.2017.2697822
25. M.R. Cunha Reis et al., Optimized techniques for driving and control of the switched reluctance motor to improve efficiency. Control Eng. Practice **90**, 1–18 (2019)
26. Z. Touati et al., Integration of switched reluctance generator in a wind energy conversion system: an overview of the state of the art and challenges. Energies **15**(13), 4743 (2022). https://doi.org/10.3390/en15134743
27. A. Arifin, I. Al-Bahadly, S.C. Mukhopadhyay, State of the art of switched reluctance generator. Energy Power Eng. **4**(6), 447–458 (2012). https://doi.org/10.4236/epe.2012.46059
28. L. Haseeb, Y. Mo, M. Mudasser, A review the research status of switched reluctance generator. Diyala J. Eng. Sci. 1–19 (2023). https://doi.org/10.24237/djes.2023.160401
29. D. Susitra, E. Annie Elisabeth Jebaseeli, S. Paramasivam, Switched reluctance generator—modeling, design, simulation, analysis and control a comprehensive review. Int. J. Comput. Appl. **1**(3), 12–25 (2010). https://doi.org/10.5120/90-189
30. M. Omar et al., Review of machine learning applications to the modeling and design optimization of switched reluctance motors. IEEE Access **10**, 130444–130468 (2022). https://doi.org/10.1109/access.2022.3229043
31. K. Vijayakumar et al., Switched reluctance motor modeling, design, simulation, and analysis: a comprehensive review. IEEE Trans. Magn. **44**(12), 4605–4617 (2008). https://doi.org/10.1109/TMAG.2008.2003334
32. L.D.S. Morais et al., Enhancement of the performance of switched reluctance generators in low wind speed conditions using advanced tracking techniques. Int. J. Energy Res. **2024**(1), 5541150 (2024)
33. G. Lefei et al., Advanced technology of switched reluctance machines in more electric aircraft: a review. IEEE Trans. Power Electron. (2025). https://doi.org/10.1109/TPEL.2024.3452171
34. B.L. Narasimharaju K. Vijay Babu, D.M. Vinod Kumar, Switched reluctance machine for off-grid rural applications: a review. IETE Tech. Rev. **33**(4), 428–440 (2016). https://doi.org/10.1080/02564602.2015.1117400

Chapter 2
Switched Reluctance Machine

2.1 Applications of Switched Reluctance Machines

The switched reluctance machine (SRM), invented by William Hannis Taylor in 1838 and patented in 1840, stands out for its simple structure, making it one of the first electrical machines developed. Despite its initial prominence, the SRM was soon surpassed by DC and AC machines because of their superior performance, particularly in terms of torque smoothness [1, 2]. Interest in SRMs resurfaced in the 1960s with advances in power electronics. The development of components such as transistors and microprocessors enabled the implementation of novel control and driving techniques, including pulse width modulation (PWM). These technological advances have contributed to significant improvements in the performance of SRM, positioning it as a competitive alternative to conventional DC and AC machines, especially in some specific applications [3, 4].

As a feature of most electrical machines, the SMR can operate as a motor and a generator. What differentiates these two modes of operation are the firing angles of the converter switches [5]. The applications of the SRM range from the aeronautical and aerospace industries to electric vehicles and wind power generation. Its application as a motor is widely distributed in the literature, mainly in high-speed applications [6]. The switched reluctance machine has a relatively simple construction, resulting in low manufacturing costs, which has led to its acceptance in various applications. Some features exhibited by the reluctance machine, such as the capability to operate under fault conditions, resilience in adverse operational environments, and high power density, have expanded its use from domestic motor applications to traction systems in the automotive industry, particularly in hybrid electric vehicles, as described in the literature [7]. Applications of the reluctance machine as a generator are significant in high-speed systems, particularly in assemblies that function as starter generators for the aerospace and automotive industries [8].

W. P. Calixto et al., *Control and Tracking Techniques for Switched Reluctance Machines*, Power Systems, https://doi.org/10.1007/978-3-031-86727-9_2

In the automotive industry, advances in vehicle technology aimed at improving comfort and safety, such as power windows, electric steering, active suspensions, and electrical systems for lighting and pumping, have rendered conventional car electrical systems inadequate due to the increased electrical energy demand. As a result, modern vehicles require more efficient, compact, and reliable energy generation systems [9]. The traditional starter motor and alternator have been replaced by an integrated starter/alternator generator connected to the combustion engine. This integrated system provides higher power and efficiency and plays a crucial role in the development of electric and hybrid vehicles [10, 11].

In the aerospace sector, conventional systems were historically powered by fluid-mechanical sources (hydraulic or pneumatic), but are now predominantly powered by electrical energy sources. Similarly to automobiles, the increasing demand for electrical energy in aircrafts has faced challenges in accommodating numerous systems. The integration of electrical energy with power electronic components is believed to have improved aircraft performance, improving reliability, fault tolerance, and power density [12, 13].

The applications of switched reluctance machines in low-speed systems are primarily focused on energy generation, particularly in wind turbines. Given the increasing integration of renewable energies into the global energy landscape, wind turbine electricity generation systems can be broadly divided into two categories: (i) Large Wind Turbines (LWTs) and (ii) Small Wind Turbines (SWTs). LWTs are typically used in wind farms and grid-connected systems, whereas SWTs are commonly used in off-grid power supply systems, such as isolated microgeneration setups. These stand-alone systems contribute positively to the provision of electrical energy in applications such as road lighting, mobile communication base towers, and individual household consumption. Even in developed countries such as the United States, SWTs are utilized to supply electricity to remote areas, supported by initiatives such as the federal government's Light for All Program [14].

Smaller wind turbines also have potential applications in developed areas where space constraints limit the expansion of existing energy supply systems. Although their industrial scale remains limited compared to that of LWT systems, its contribution to energy generation is equally significant. Ongoing research into the use of switched reluctance generators for wind energy applications suggests that they could be viable alternatives to conventional machines. Current technologies for wind energy generation include permanent magnet generators, doubly-fed induction generators, synchronous generators, and squirrel cage induction generators. Each of these generators presents specific challenges [15].

For instance, squirrel-cage induction generators operate at a constant speed and require capacitors with high capacitance to achieve a nearly unity power factor. Permanent magnet generators face high costs and the risk of material demagnetization, in addition to operating at high torque, which is disadvantageous for low-speed applications. Although induction machines are known for their resilience and low maintenance, they produce electrical energy with low efficiency. The doubly-fed induction generator raises safety concerns during fault conditions. The growing interest in switched reluctance machines as generators has spurred comprehensive research in areas such as control strategies, optimization methods, machine design improvements, structural improvements, and power converter topologies [16, 17].

In addition to wind generation, other machine applications include electric and hybrid vehicles, automotive auxiliary systems, aeronautics (such as actuators and emergency generators), industry (machine tools, pumps, and compressors), household devices (vacuum cleaners, washing machines, and fans), medical sector (imaging equipment and life support), distributed generation (hydroelectric plants and marine turbines), robotics, and transport (trains, meters, and elevators). This versatility makes SRMs an attractive choice in scenarios that require high reliability and efficiency [8, 15, 18–20].

2.2 Construction Characteristics of Switched Reluctance Machines

From a construction perspective, the switched reluctance machine is characterized by salient poles on both the rotor and the stator. These poles extend into the air gap of the magnetic circuit, which is composed primarily of the magnetic material in the stator and rotor. Both components consist of stacked, insulated sheets of silicon steel (typically non-oriented grain) to minimize eddy currents. Figure 2.1 shows the lamination packages of the rotor and the stator of a switched reluctance machine with a configuration 6×4, indicating a stator with six poles and a rotor with four poles [21].

The number of stator and rotor poles is designed to prevent regions with zero torque, resulting in configurations with symmetrically and evenly spaced poles [22]. The machine operates effectively with unidirectional rectangular current pulses applied to its phase coils [4]. The switched reluctance machine inherently operates by consuming (in motor mode) or generating (in generator mode) pulsed direct current. Proper design of the switched reluctance machine and its corresponding power supply circuit can achieve higher efficiency than that of an induction machine

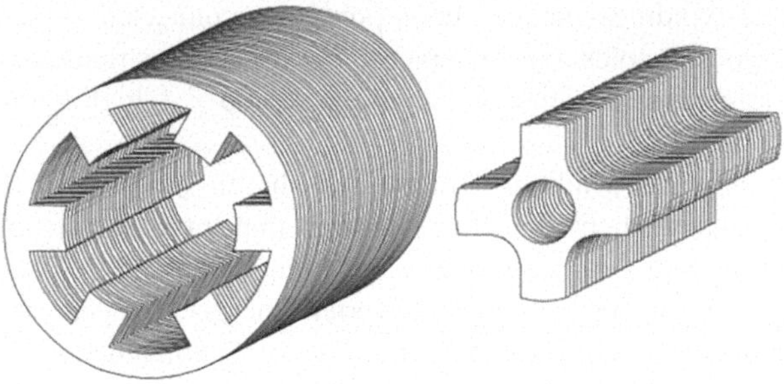

Fig. 2.1 Lamination packages of the stator and rotor of the switched reluctance machine

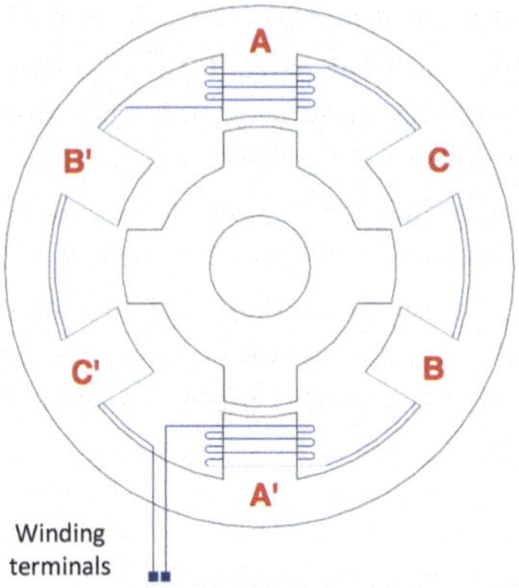

Fig. 2.2 Cross section of the switched reluctance machine 6 × 4

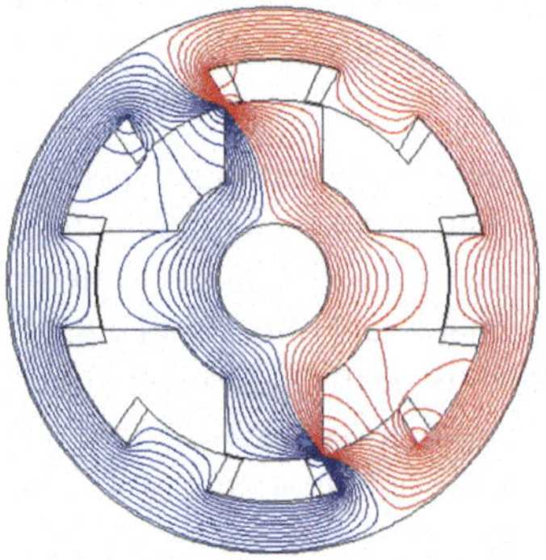

Fig. 2.3 Flux lines of the magnetic circuit formed as a result of energizing one of the phases of the switched reluctance machine

under variable-speed conditions, with the added advantage of simpler construction for both the machine and its converter [23].

Typical configurations include three-phase machines with 6 stator poles and 4 rotor poles, as shown in Fig. 2.1, three-phase machines with a 12 × 8 topology, four-phase machines with a 8 × 6 topology, and four-phase machines with a 16 × 12 topology [24]. The phase coils of the switched reluctance machine are concentrated exclusively in the stator, allowing the rotor to remain free of windings, magnets, or brushes. The windings on the stator poles are connected in series with the diametrically opposite poles to form a phase. Figure 2.2 illustrates the cross section of a switched reluctance machine with configuration 6 × 4, highlighting one of the three phases [25].

In Fig. 2.2, only the winding of phase A is highlighted, while the other two phases (B and C) have their respective windings that form the machine coils. This construction ensures phase independence, with each phase individually energized by an external source through a power converter. Energizing a phase establishes the magnetic circuit. Figure 2.3 shows the flux lines of the magnetic circuit resulting from the energization of one phase.

When a phase winding is energized, the corresponding stator pole acts as an electromagnet, attracting the rotor toward the excited phase. Torque production in the machine results from the rotor's tendency to align with the stator pole to establish a magnetic circuit with minimum reluctance (or maximum inductance) [4, 22].

2.3 Working Principle of Switched Reluctance Machines

The operation of the switched resistance machine is based on the individual energization of each phase, which must be synchronized with the instantaneous position of the rotor θ. This synchronization is essential due to the inductance profile of each phase and the torque production equation T. As the rotor's angular position θ increases from the reference axis, a rotor pole approaches a stator pole, reducing the air gap of the magnetic circuit and increasing the inductance of the corresponding stator pole coil. This behavior is depicted in Fig. 2.4.

In Fig. 2.4, the parameters L_a and L_u represent the maximum and minimum inductance of a machine phase, respectively. L_a corresponds to the inductance when the magnetic circuit has the minimum air gap, which occurs when the rotor and stator poles are fully aligned. In contrast, L_u represents the low inductance value when the rotor and stator poles are misaligned, resulting in a large air gap in the magnetic circuit of the respective phase. According to the per-phase inductance behavior, the magnetic circuit equation provided indicates that the torque produced by this circuit is given by citearaujo2021efficiency:

$$T = \frac{1}{2} \cdot \frac{dL}{d\theta} \cdot i^2 \tag{2.1}$$

In expression (2.1), it can be observed that: (i) the torque T is independent of the direction of current flow in the coil, and ii)the sign of the torque depends on the timing of the electric current pulse. This implies that the torque produced by the machine is positive if the current pulse is applied during the positive rate of change of inductance with respect to the angular position, i.e., when $\frac{dL}{d\theta} > 0$. Consequently, the machine produces positive torque and operates as a motor if the phase coils are

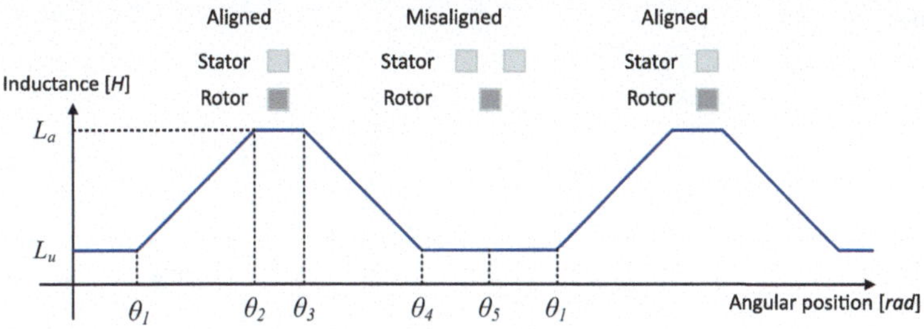

Fig. 2.4 Behavior of the inductance of one of the phases as a function of the angular position θ

energized during the interval of increasing inductance, specifically between θ_1 and θ_2 in Fig. 2.4.

In contrast, if the machine's rotor is driven by external mechanical energy and current is applied to the phase windings during the interval of negative inductance variation, i.e., between θ_3 and θ_4 in Fig. 2.4, where $\frac{dL}{d\theta} < 0$, the torque produced by the machine is negative. As a result, the generated torque opposes the applied mechanical energy, allowing this energy to be transferred to the load, thus causing the machine to operate as a generator. The operating principle of the machine requires specific requirements for its drive system, including: (i) a position sensor to determine the instantaneous value of θ for precise coil energization timing, and (ii) a power converter capable of applying current pulses to the coils based on the θ value provided by the sensor. These requirements have also been the focus of scientific research, leading to studies on sensorless drive methods and various power converter topologies for coil energization [3].

2.4 Power Converter

A power converter is essential for the appropriate energization of the coils in the switched resistance machine, regardless of its mode of operation, motor, or generator. The converter most commonly used to drive the switched reluctance machine is the Asymmetric Half-Bridge (AHB) converter, as shown in Fig. 2.5.

The diagram in Fig. 2.5 shows the drive circuit for one of the machine's phases, specifically phase A (F_A). Each additional phase requires a similar configuration with a pair of diodes and switches. The switches enable individual control of each machine coil. The AHB topology depicted in Fig. 2.5 is suitable for motor operation, where current pulses must be applied during the positive rate of change in the coil's inductance to produce positive torque. In this mode, the current flows through the switches and the coil. When the switches are open, the back electromotive force generated in the coil allows current to continue flowing through the coil in the

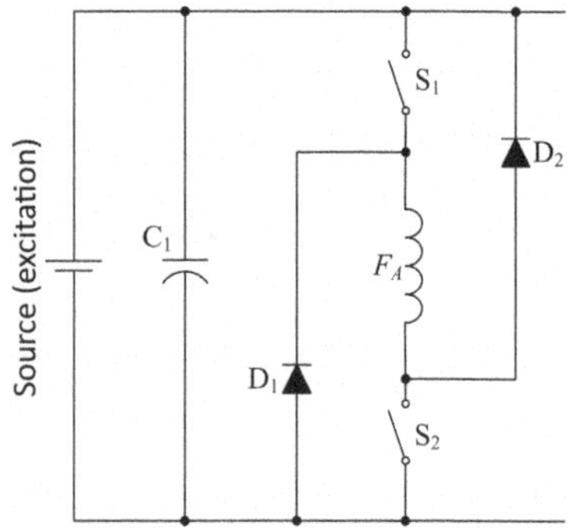

Fig. 2.5 *AHB* converter schematics

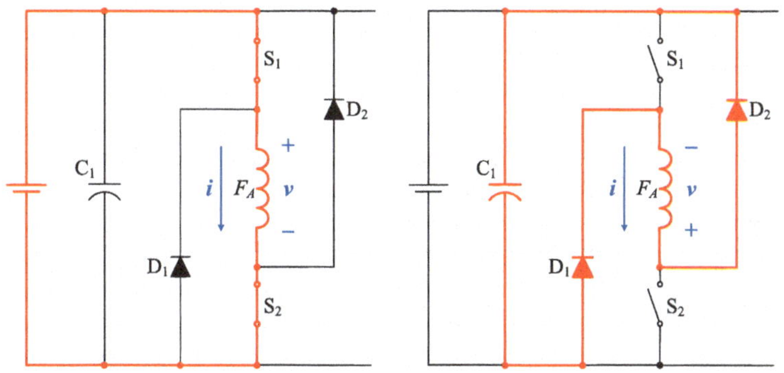

Fig. 2.6 Steps for activating one of the machine's phases

same direction via the diodes. A capacitor is necessary to store the energy released from the coil. Figure 2.6 illustrates both stages of the driving process for one of the machine's coils.

There are two stages involved in driving a machine phase: the coil magnetization stage and the regeneration stage. The waveforms shown in Fig. 2.7 are idealized and do not represent the actual behavior of these quantities. In the graphs in Fig. 2.7, the parameter C_R denotes the rotor cycle, corresponding to the period over which the stator quantities repeat with respect to θ. For a switched reluctance machine with a configuration $N_e \times N_r$, where N_e is the number of stator poles, and N_r is the number of rotor poles, the rotor cycle is defined as

$$C_R = \frac{2 \cdot \pi}{N_r} \tag{2.2}$$

For a switched reluctance machine with a topology of 6 × 4, the rotor cycle is 90 degrees. In practical applications, switches S_1 and S_2 of the AHB converter shown in Fig. 2.5 are power semiconductor switches. These switches facilitate rapid opening and closing within short time intervals, allowing the conditioning of electrical energy in the coils and enabling high voltage and current levels. The same converter configuration can be used for generator operation, as shown in Fig. 2.8.

In the circuit shown in Fig. 2.8, the three phases of the machine 6 × 4 are represented as F_A, F_B, and F_C. The main difference between this circuit and the converter in Fig. 2.5 is that the diodes are connected to a separate bus linked to the load, rather than the same excitation bus. Thus, the regeneration stage becomes the generation stage, as the electromotive force behind the coils, associated with the closing of switches S_1 to S_6, is directed to the load instead of the capacitor C_1. The capacitor C_2 serves to smooth the oscillatory nature of the load voltage, providing DC power at the generator output.

The switching instants θ_{ON} and θ_{OFF}, shown in Fig. 2.7 and defined as the turn-on and turn-off angles of the phases, directly influence the machine's output power, both in mechanical power during motor operation and electrical power during

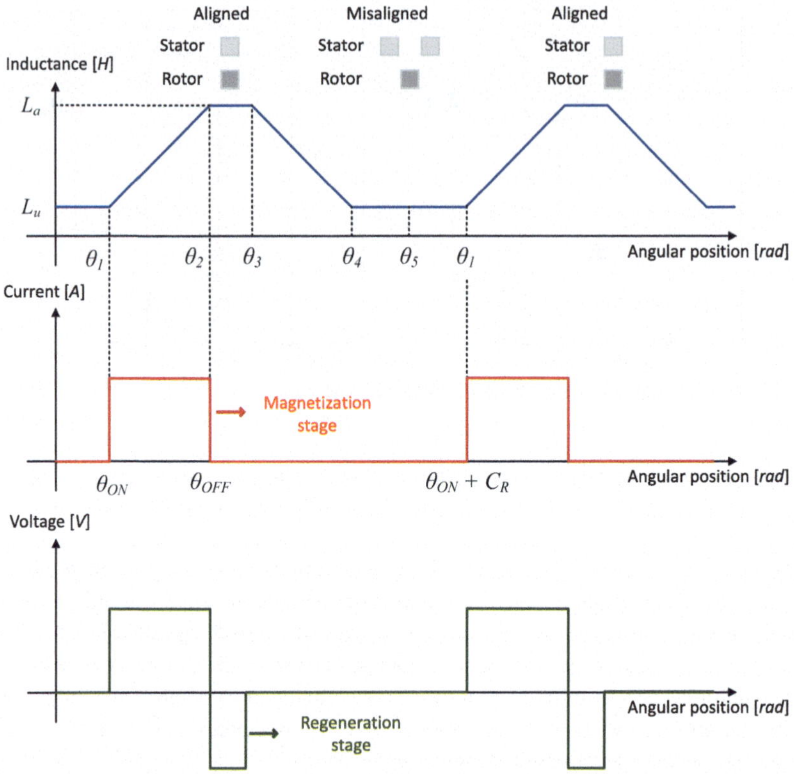

Fig. 2.7 Hypothetical illustration of current and voltage behavior in the coil for the drive stages

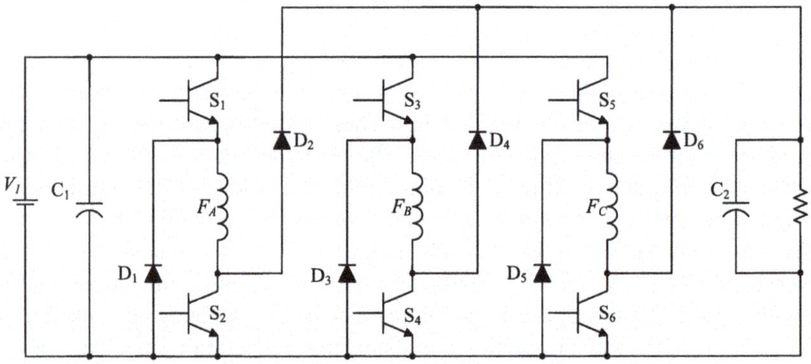

Fig. 2.8 Schematic of the AHB converter to drive the machine as a generator

generator operation. The SRM drive system is relatively complex because the torque is generated only when the inductance of a given phase increases as a function of the rotor's angular position, as described in (2.7). Therefore, continuous monitoring of

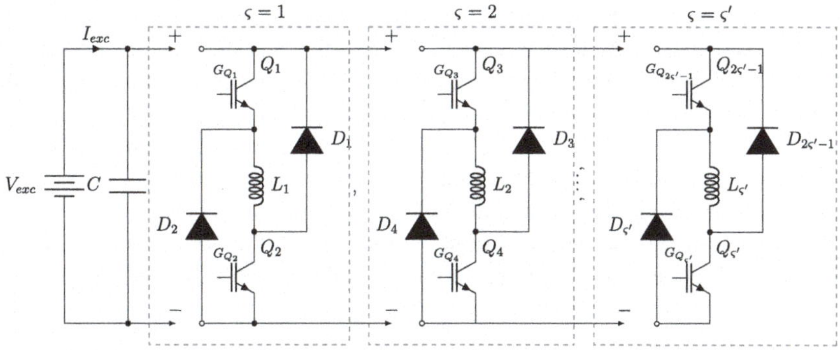

Fig. 2.9 AHB converter for driving the SRM in Phase ς

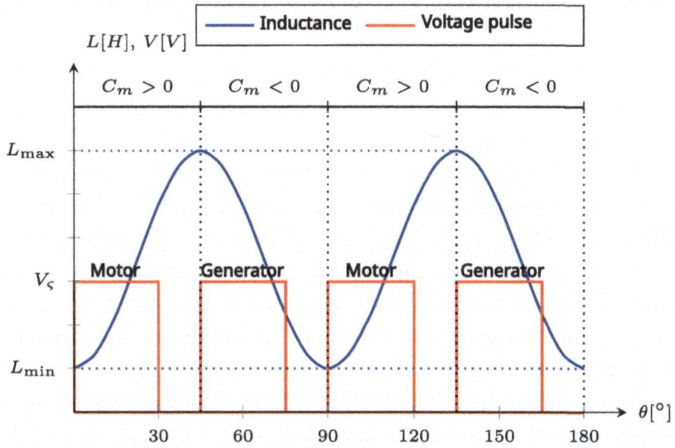

Fig. 2.10 Voltage pulse applied to phase ς of the switched reluctance machine during the motorization and generation period

the rotor's instantaneous position is necessary to ensure that the drive occurs within the region of increasing inductance. This is typically accomplished using optical sensors or encoders attached to the SRM shaft, which detect the rotor's position to switch voltage pulses on each phase at the correct time. The most commonly used converter for SRM drives is the asymmetric half-bridge, also known as the half-bridge converter, shown in Fig. 2.9 [3].

The pulses from the optical sensor, interpreted by the switching logic, are directed to the triggers G_Q of the switches Q to activate the SRM coils at the appropriate times. The drive process involves applying voltage pulses to G_Q $[V]$ during the phase's increasing inductance for motor operation. Figure 2.10 illustrates the voltage pulse applied to phase ς of the SRM during periods of increasing and decreasing inductance. When the switches are activated at positions where the inductance derivative is positive, the machine produces a positive torque, functioning as a motor. In contrast, when activated at positions where the inductance derivative is negative, the torque produced is negative and the machine operates as a generator, as described in (2.7) [3, 4]. In Fig. 2.10, it can be seen that applying the command pulse to the triggers of the switches $Q_{2\varsigma-1}$ and Q_{ς} connects phase F_{ς} to the direct current (DC) bus of the half-bridge converter. The excitation voltage source $V_{exc}[V]$ must be designed to supply direct current $I_{exc}[A]$ to the coils during the intervals defined by the switching angles $\theta_{on}[°]$ and $\theta_{off}[°]$ [4].

The operation of the SRM, similar to that of other electric motors, can be controlled using techniques tailored to the requirements of industrial, automotive, aerospace, and other sectors.

2.5 Switched Reluctance Generator Excitation Strategies

The requirement for continuous rotor position monitoring contributes to the complexity of driving the SRM. Due to this characteristic, applications involving switched reluctance machines have advanced significantly since the 1970s, supported by the development of power electronics and microprocessors, which enhance SRM performance. There are two primary methods for stimulating the SRM, as illustrated in Fig. 2.11. One method involves an independent excitation source with a fixed voltage, while the other employs self-excited operation. For both strategies, the SRM power converter must be able to magnetize the phase winding at the correct instant, regardless of the mode of operation, as torque production depends on the rotor position [26].

The converter commonly used to drive the SRM, regardless of the excitation method, is the Asymmetric Half-Bridge (AHB) converter [3, 4, 26]. Figure 2.8

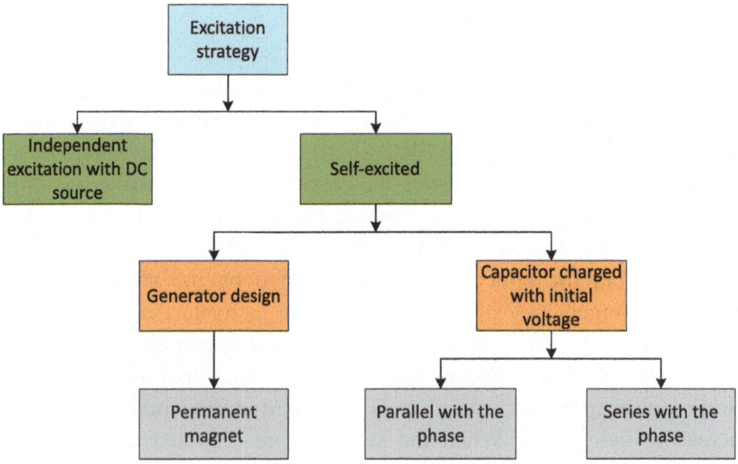

Fig. 2.11 Main excitation methods for the switched reluctance generator

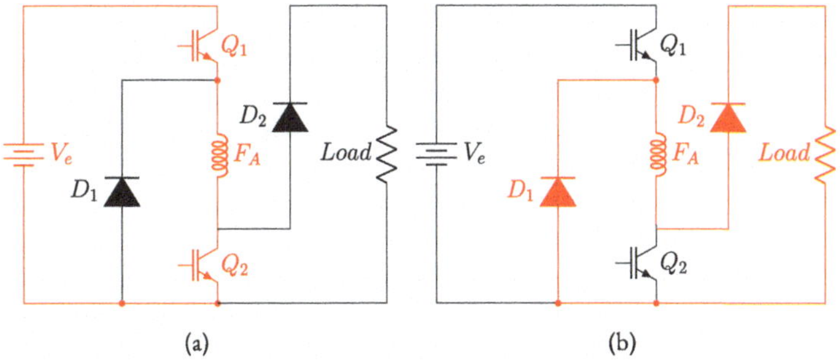

Fig. 2.12 Operating stages of the AHB converter with independent excitation source: (**a**) excitation stage and (**b**) generation stage

shows this converter configured to drive the switched resistance generator using an independent excitation source. Each phase of the converter consists of two switches Q in series with the coil and two diodes D, allowing individual current control for each phase. This configuration allows one phase to be magnetized while another is still demagnetizing. Figure 2.12 illustrates the operating cycle of one phase of the converter.

The first stage of operation of the AHB converter is the excitation stage, during which both switches Q_1 and Q_2 are closed, connecting the voltage source V_e to the respective phase coil. The current flows from the positive terminal of V_e, through the switch S_1, the phase coil, the switch S_2, and then to the negative terminal of V_e. Following the excitation (magnetization) stage, the second operating stage, the generation stage, occurs. In this stage, both switches Q_1 and Q_2 are opened, and the back electromotive force generated in the coil allows the current to flow through the coil and the diodes D_1 and D_2. Each phase of the drive system involves two main stages. In addition, some drive strategies include an intermediate stage, called the free-wheeling stage, which occurs after the excitation stage. When switch Q_1 is opened, the current is circulating through switch Q_2 and diode D_1. During this stage, the concatenated flux remains constant, while the current continues to increase due to the electromotive force of the coil. This stage is introduced when zero voltage is desired [2].

The SRG does not possess self-excitation capability; however, permanent magnets can be added to the machine's stator to provide magnetic flux for excitation. This strategy has several drawbacks, including: (i) increased manufacturing costs, (ii) dependence on speed to generate sufficient flux for excitation, and (iii) complications in the manufacturing process. An alternative approach for operating the SRG in self-excited mode without altering the machine's construction is to add capacitors in series or parallel with the phase winding and load. The AHB converter is used in this configuration, and Fig. 2.13 illustrates the schematic of the AHB converter to drive the machine as a generator in self-excited mode with a capacitor in parallel with the load.

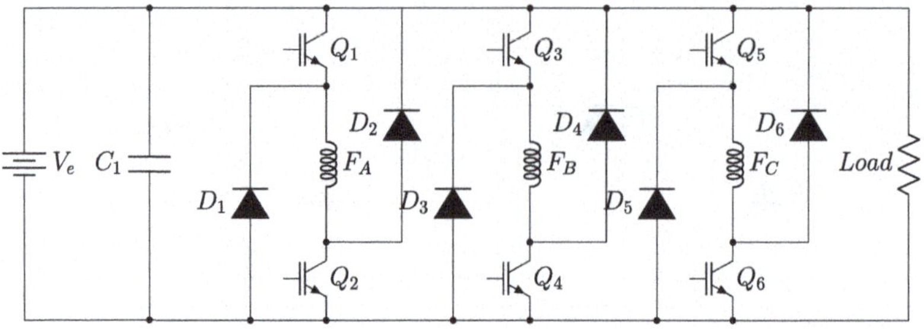

Fig. 2.13 Schematic of the AHB converter to drive the machine as a self-excited generator

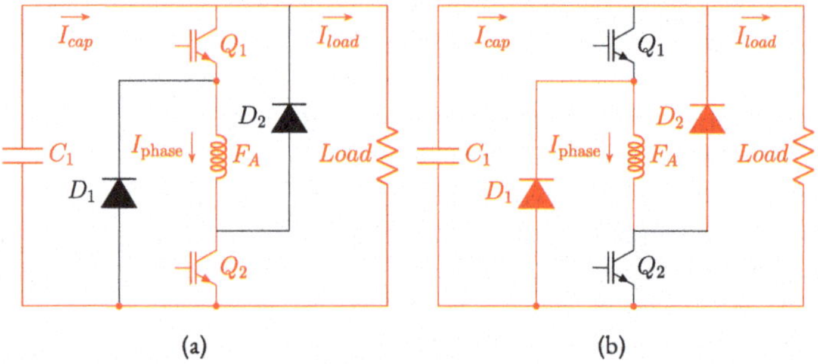

Fig. 2.14 Operating steps of the AHB converter with self-excitation: (**a**) excitation step and (**b**) generation step

In addition to the capacitor connected in parallel with the load, another difference compared to independent excitation operation is that the excitation bus also receives the generated energy. In this process, the initial charging of the capacitor no longer requires an external excitation source, as the capacitor begins to excite the machine's phases. Figure 2.14 illustrates the excitation and generation stages. The current of the capacitor, I_{cap}, splits into two paths: one part is used for phase excitation, I_{phase}, while the other supplies the load, I_{load}. Once the excitation stage starts, the phase current increases as a result of the excitation current and the back electromotive force. When switches Q_1 and Q_2 open, the electromotive force on the back allows the current to flow through the diodes D_1 and D_2 [4].

2.6 Mathematical Model of the Switched Reluctance Machine

The switched reluctance motor has relatively simple structural characteristics compared to various other types of electric motor. It is typically constructed using nonoriented grain silicon steel sheets stacked to form the magnetic core [4]. The stator and rotor poles are salient and diametrically opposed, with the windings of the

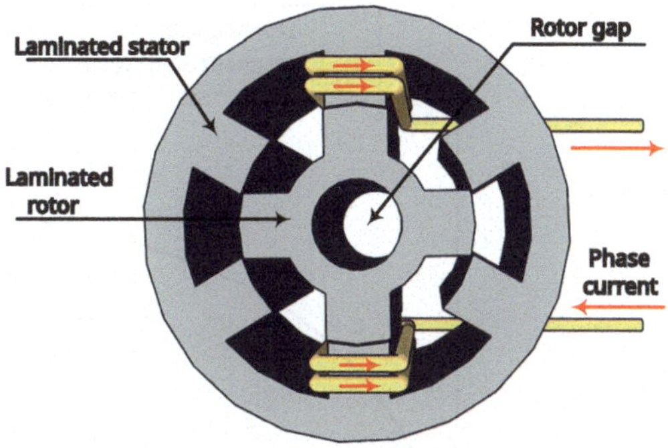

Fig. 2.15 Hypothetical representation of the switched reluctance motor with characteristic 6 × 4 with single-phase winding

machine's phases located on the stator, while the rotor is unwound. These features contribute to the structural simplicity of the machine, positively affecting manufacturing costs. In addition, the absence of friction between moving parts enables high-speed operation without significant issues, making this type of electric machine suited for automotive and aerospace applications [27]. Figure 2.15 illustrates the SRM with a configuration 6 × 4, showing the laminated rotor and stator assembly and the winding of one phase.

Switched reluctance machines can have various stator and rotor pole configurations, which are subjects of extensive research in the scientific community [4, 27]. The mathematical model of the SRM is based on fundamental equations that describe its electrical and mechanical characteristics and is crucial to developing the computational model used in simulations. The simulator allows for the analysis of SRM behavior, facilitating the study of new drive and control techniques, among other aspects [28, 29]. The modeling that defines the SRM based on phase voltage is represented by:

$$V_\varsigma = R_\varsigma \cdot i_\varsigma + \frac{d\Psi_\varsigma(t)}{dt} \tag{2.3}$$

where R $[\Omega]$ and i $[A]$ represent the resistance and current in phase ς, respectively, with $1 \leq \varsigma \leq \varsigma'$ over integer intervals. ς' denotes the total number of phases in the SRM under study, and Ψ $[Wb]$ is the concatenated flux produced by winding ς [30]. The concatenated flux is determined by the product of self- and mutual inductance and their respective currents. For the 6 × 4 SRM, the mutual inductance between phases is generally minimal and can be neglected. Experimental studies by various authors have shown that this value is less than 1% compared to the self-inductance of the machine [31]. Therefore, by neglecting mutual inductance between phase windings, Ψ can be expressed as $\Psi = L_\varsigma(\theta, i) \cdot i_\varsigma$, where L $[H]$ is the inductance of the winding ς [30]. By substituting L into (2.3):

$$V_\varsigma = R_\varsigma \cdot i_\varsigma + L_\varsigma(\theta, i) \cdot \frac{di_\varsigma(t)}{dt} + i_\varsigma \cdot \omega \cdot \frac{\partial L_\varsigma(\theta, i)}{\partial \theta} \tag{2.4}$$

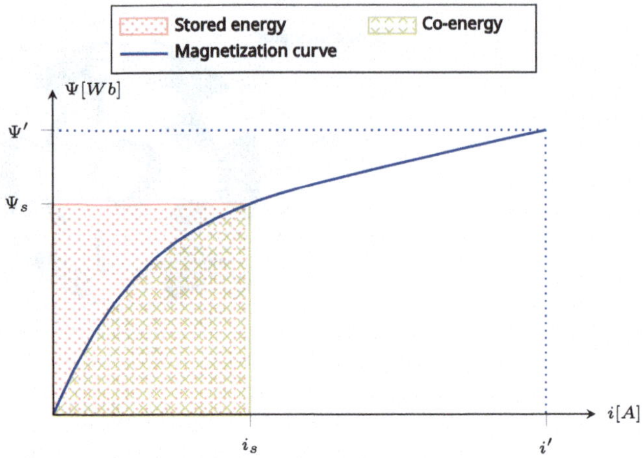

Fig. 2.16 Magnetization curve of the switched reluctance motor

where ω [rad/s] is the speed, and θ [rad] represents the rotor position. When one of the machine's phases is excited, the magnetic material follows a magnetization curve that depends on the applied current. For any given angular position θ, the co-energy W' is defined as the area under the magnetization curve, as shown in Fig. 2.16 and is expressed as:

$$W' = \int_0^{i_s} \Psi(\theta, i) \cdot di \tag{2.5}$$

where Ψ_s is the flux at which the magnetic material reaches saturation, Ψ_s' is the flux under the machine design limit conditions, i_s is the current applied to the coil that results in Ψ_s, and i' is the machine design limit current. The general expression for the mechanical torque produced by the machine is given by:

$$C_m = \frac{\partial W'(\theta, i)}{\partial \theta} \tag{2.6}$$

Substituting (2.5) into (2.6) and applying the defined derivative yields:

$$C_m = \frac{i_1^2}{2} \cdot \frac{\partial L_1(\theta, i)}{\partial \theta} + \frac{i_2^2}{2} \cdot \frac{\partial L_2(\theta, i)}{\partial \theta} + \cdots + \frac{i_{\varsigma'}^2}{2} \cdot \frac{\partial L_{\varsigma'}(\theta, i)}{\partial \theta} \tag{2.7}$$

To define the mechanical behavior of the motor, the forces must be formulated to achieve equilibrium. The electrical energy supplied to the motor generates mechanical torque C_m [N · m] and counters the load torque C_c [N · m], taking into account the effects of inertia J [(N · m · s^2)/rad] and friction D [(N · m · s)/rad] as mechanical losses [30]. The energy efficiency η is defined as the ratio of the mechanical power produced P_m to the electrical power consumed P_e. The expressions that describe the mechanical behavior of the motor and the energy efficiency η are given by:

$$C_m = C_c + D \cdot \omega + J \cdot \frac{d\omega(t)}{dt} \tag{2.8}$$

$$\eta = \frac{P_m}{P_e} = \frac{C_m \cdot \omega}{V_{exc} \cdot I_{exc}} \tag{2.9}$$

where $V_{exc}[V]$ is the excitation voltage of the direct current (DC) bus and $I_{exc}[A]$ is the excitation current of the DC bus. Considering the contribution of the three machine phases and the mechanical equation (2.8), the mathematical model of the SRM can be represented by the state equation, expressed as:

$$[V]_{[(\varsigma'+2),1]} = [R]_{[(\varsigma'+2),(\varsigma'+2)]} \cdot [I]_{[(\varsigma'+2),1]} + [L]_{[(\varsigma'+2),(\varsigma'+2)]} \cdot [\dot{I}]_{[(\varsigma'+2),1]} \tag{2.10}$$

In (2.10), the subscripts next to the variables V, R, L, and I indicate the dimensions of the matrices they represent, defined as:

$$\begin{bmatrix} V_1 \\ V_2 \\ \vdots \\ V_{\varsigma'} \\ C_c \\ 0 \end{bmatrix} = \begin{bmatrix} R_1 & 0 & 0 & 0 & 0 & 0 \\ 0 & R_2 & 0 & 0 & 0 & 0 \\ 0 & 0 & \ddots & 0 & 0 & 0 \\ 0 & 0 & 0 & R_{\varsigma'} & 0 & 0 \\ b_1 & b_2 & \cdots & b_{\varsigma'} & -D & 0 \\ 0 & 0 & 0 & 0 & -1 & 0 \end{bmatrix} \cdot \begin{bmatrix} i_1 \\ i_2 \\ \vdots \\ i_{\varsigma'} \\ \omega \\ \theta \end{bmatrix} + \begin{bmatrix} L_1 & 0 & 0 & 0 & 0 & 2b_1 \\ 0 & L_2 & 0 & 0 & 0 & 2b_2 \\ 0 & 0 & \ddots & 0 & 0 & \vdots \\ 0 & 0 & 0 & L_{\varsigma'} & 0 & 2b_{\varsigma'} \\ 0 & 0 & 0 & 0 & -J & 0 \\ 0 & 0 & 0 & 0 & 0 & 1 \end{bmatrix} \cdot \begin{bmatrix} \dot{i}_1 \\ \dot{i}_2 \\ \vdots \\ \dot{i}_{\varsigma'} \\ \dot{\omega} \\ \dot{\theta} \end{bmatrix} \tag{2.11}$$

where,

$$b_\varsigma = \frac{i_\varsigma}{2} \cdot \frac{\partial L_\varsigma(\theta, i)}{\partial \theta} \tag{2.12}$$

where $1 \leq \varsigma \leq \varsigma'$ over integer intervals, with ς' representing the number of phases in the SRM under study. This allows for the derivation of a state equation model that represents each phase of the machine when voltage is applied, under conditions where the inductance produces positive torque, as described in (2.7), for motor operation, given by:

$$[\dot{I}]_{[(\varsigma'+2),1]} = [L]^{-1}_{[(\varsigma'+2),(\varsigma'+2)]} \cdot [V]_{[(\varsigma'+2),1]} - [L]^{-1}_{[(\varsigma'+2),(\varsigma'+2)]} \cdot [R]_{[(\varsigma'+2),(\varsigma'+2)]} \cdot [I]_{[(\varsigma'+2),1]} \qquad (2.13)$$

The mathematical model of the switched reluctance generator (SRG) is typically developed to design a computational system for generator simulation. Simulating this type of machine is more complex than simulating conventional AC and DC machines due to several contributing factors [22]: (i) the nonlinearity of the magnetic material's $B \times H$ characteristic; (ii) the dependence of the concatenated flux and inductance on both the rotor's angular position θ and the current amplitude. Although this dependence also exists in other machines, it can be mitigated in those cases through trigonometric transformations, which is not feasible for the SRG; and (iii) the presence of a single power supply. Consequently, the concatenated flux λ and the inductance depend simultaneously on two parameters:

$$\lambda(\theta, i) = L(\theta, i) \cdot i \qquad (2.14)$$

Figure 2.17 illustrates the dependence of the concatenated flux on current and angular position. In this figure, the various curves represent the concatenated flux as a function of the coil current for different angular positions θ, with the upper curve corresponding to the flux at maximum inductance when the rotor and stator poles

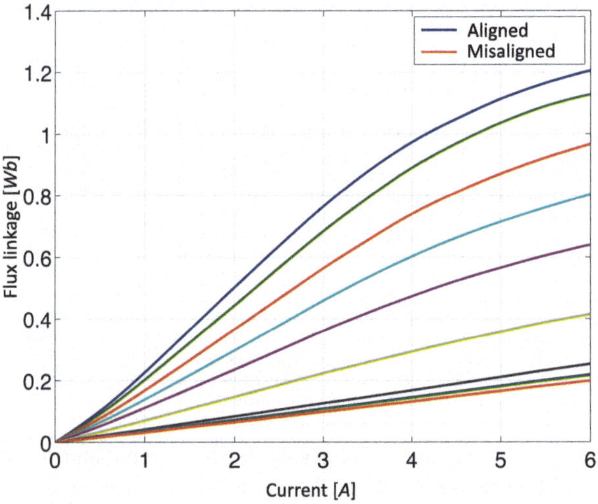

Fig. 2.17 Behavior of the flux linkage as a function of the current and the angular position of the rotor

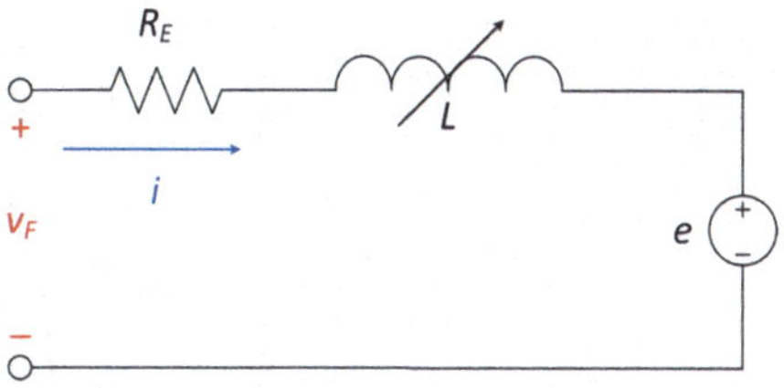

Fig. 2.18 Equivalent electrical circuit of one phase of SRG

are aligned, and the lower curve corresponding to the flux at the fully unaligned position. Therefore, the mathematical model must account for these factors.

Thus, considering the *RL* circuit formed by the single-phase winding, as illustrated in Fig. 2.18:

Neglecting the mutual inductance between the coils, the expression for the circuit shown in Fig. 2.18 is:

$$v_F = R_E \cdot i + \frac{d\lambda(\theta, i)}{dt} \tag{2.15}$$

where v_F is the DC voltage applied to the phase coil, R_E is the resistance of the phase winding, L is the phase inductance, and e is the counter electromotive force. Thus, (2.15) becomes:

$$v_F = R_E \cdot i + L(\theta, i) \cdot \frac{di}{dt} + e \tag{2.16}$$

The counter electromotive force can be expressed separately as:

$$e = \omega \cdot i \cdot \frac{dL(\theta, i)}{d\theta} \tag{2.17}$$

where ω is the angular velocity of the rotor, defined as:

$$\omega = \frac{d\theta}{dt} \tag{2.18}$$

The mechanical expression of the machine can be derived from the dynamic analysis of its operation. For generator operation, the primary machine must apply the external mechanical torque T_M. The force balance is represented by:

$$T_M = T_{EM} + D \cdot \omega + J \cdot \frac{d\omega}{dt} \tag{2.19}$$

where T_{EM} is the electromagnetic torque produced by the machine, D is the viscous friction coefficient, and J is the moment of inertia. To determine the value of the electromagnetic torque T_{EM}, the typical magnetization curve for the reference angular position θ_{REF}, shown in Fig. 2.19, must be considered.

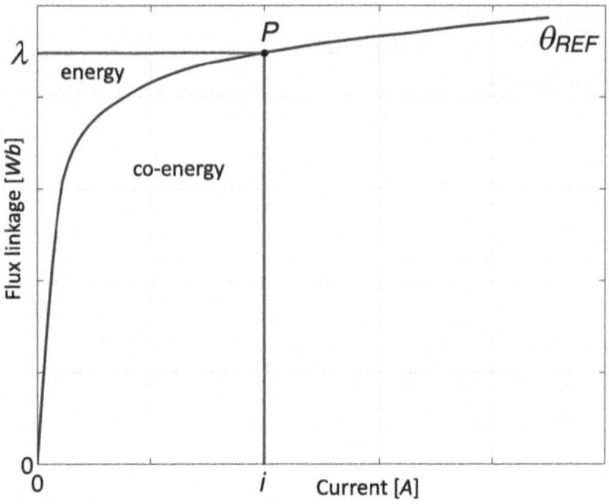

Fig. 2.19 Typical magnetization curve for an SRG phase

The co-energy W_{CO} of a given phase, for the position θ, is given by:

$$W_{CO}(\theta, i) = \int_0^i \lambda(\theta, i) \cdot di \tag{2.20}$$

where λ represents the flux linkage of the phase coil. Since this expression applies to each of the three phases of the 6×4 SRG, the total co-energy W_{CO_T} is:

$$W_{CO_T}(\theta, i) = W_{CO_1} + W_{CO_2} + W_{CO_3} \tag{2.21}$$

The instantaneous electromagnetic torque can be expressed as:

$$T_{EM} = \frac{\partial W_{CO}(\theta, i)}{\partial \theta} \tag{2.22}$$

The expansion of this expression to the three phases results in:

$$T_{EM} = \frac{\partial W_{CO_1}(\theta, i)}{\partial \theta} + \frac{\partial W_{CO_2}(\theta, i)}{\partial \theta} + \frac{\partial W_{CO_3}(\theta, i)}{\partial \theta} \tag{2.23}$$

Replacement of (2.23) in (2.18):

$$T_M = (\frac{\partial W_{CO_1}(\theta, i)}{\partial \theta} + \frac{\partial W_{CO_2}(\theta, i)}{\partial \theta} + \frac{\partial W_{CO_3}(\theta, i)}{\partial \theta}) + B \cdot \omega + J \cdot \frac{d\omega}{dt} \tag{2.24}$$

Grouping of the electrical (2.16) and mechanical (2.24) expressions and rewriting them in matrix form:

$$\begin{bmatrix} v_A \\ v_B \\ v_C \\ T_M \\ 0 \end{bmatrix} = \begin{bmatrix} R_A & 0 & 0 & 0 & 0 \\ 0 & R_B & 0 & 0 & 0 \\ 0 & 0 & R_C & 0 & 0 \\ r_1 & r_2 & r_3 & D & 0 \\ 0 & 0 & 0 & -1 & 0 \end{bmatrix} \cdot \begin{bmatrix} i_A \\ i_B \\ i_C \\ \omega \\ \theta \end{bmatrix} +$$

$$+ \begin{bmatrix} L_A(\theta, i) & 0 & 0 & 0 & i_A \cdot \dfrac{\partial L_A(\theta, i)}{d\theta} \\ 0 & L_B(\theta, i) & 0 & 0 & i_B \cdot \dfrac{\partial L_B(\theta, i)}{d\theta} \\ 0 & 0 & L_c(\theta, i) & 0 & i_C \cdot \dfrac{\partial L_C(\theta, i)}{d\theta} \\ 0 & 0 & 0 & J & 0 \\ 0 & 0 & 0 & 0 & 1 \end{bmatrix} \cdot \begin{bmatrix} \dot{i}_A \\ \dot{i}_B \\ \dot{i}_C \\ \dot{\omega} \\ \dot{\theta} \end{bmatrix} \tag{2.25}$$

where variables r_1, r_2, and r_3 are defined as:

$$r_1 = \frac{i_A}{2} \cdot \frac{\partial L_A(\theta, i)}{\partial \theta} \tag{2.26}$$

$$r_2 = \frac{i_B}{2} \cdot \frac{\partial L_B(\theta, i)}{\partial \theta} \tag{2.27}$$

$$r_3 = \frac{i_C}{2} \cdot \frac{\partial L_C(\theta, i)}{\partial \theta} \tag{2.28}$$

Assigning the terms $[v_F]$, $[R_E]$, $[I]$, $[L]$, and $[\dot{I}]$ to each matrix in the order in which they appear in expression (2.25), we obtain:

$$[v_F] = [R_E] \cdot [I] + [L] \cdot [\dot{I}] \tag{2.29}$$

With expression (2.29), the machine state equations can be written as:

$$[\dot{I}] = [L^{-1}] \cdot [v_F] - [L^{-1}] \cdot [R_E] \cdot [I] \tag{2.30}$$

The expression (2.30) fully describes the state of each phase of the machine at any given moment. The coil resistance and inductance values can be measured directly at the machine terminals or determined from the construction parameters using numerical techniques such as the finite element method. The expression (2.30) represents the differential equation of the system that relates the input v_F to the corresponding output i. The solution to (2.30) is obtained using iterative numerical methods.

2.7 Computational Model of the Switched Reluctance Machine

With the mathematical model established, a computational system can be developed to simulate the electrical and mechanical behavior of the SRG. Analyzing these behaviors facilitates the study of optimization, control, and drive techniques to enhance machine performance. Figure 2.20 illustrates the basic diagram used for SRG simulation.

The computational simulation of the switched reluctance machine, whether for motor or generator operation, follows the steps outlined in the diagram in Fig. 2.20. Initially, parameters such as the resistance of the stator coils and mechanical properties such as friction and inertia are input into the system. Next, the mathematical expression in (2.30) is solved on the basis of the coil voltages, providing the power converter circuit with the instantaneous phase current values, as the machine coils are modeled as current sources. This converter (illustrated in Fig. 2.5) is controlled by the switching provided by the Switching Logic block in Fig. 2.20. Finally, the results are updated, yielding instantaneous values of the generator's electrical and mechanical quantities for performance verification.

However, the operation of the computational model is heavily based on the block described as the Inductance Model, as indicated in Fig. 2.20. The manner in which coil inductance and its rate of variation are represented directly impacts the simulation and, most importantly, the accuracy of the values produced by the model compared to the experimental data. The simplest approach to represent the SRG inductance is the linear model. This model is termed "linear" because it assumes that inductance behaves as a sinusoidal, simplifying the rate of variation. Additionally, this representation does not account for magnetic saturation in the circuit. Given that

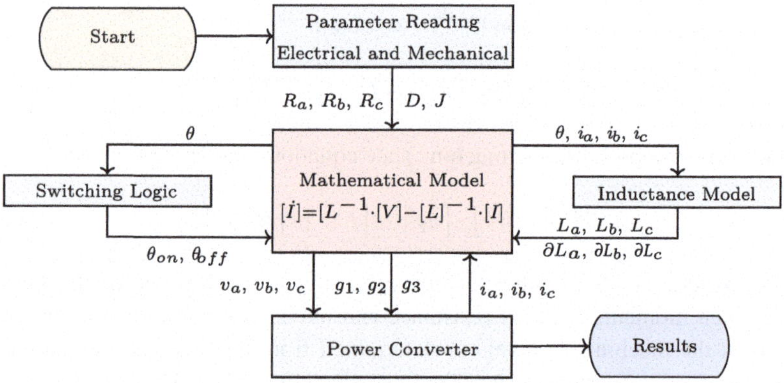

Fig. 2.20 Flowchart of switched reluctance generator simulation

the stator poles are distributed equidistantly at 60° intervals, the coil inductance can be represented by:

$$L_A = A + B \cdot \cos(4 \cdot \theta) \tag{2.31}$$

$$L_B = A + B \cdot \cos[4 \cdot (\theta - 60°)] \tag{2.32}$$

$$L_C = A + B \cdot \cos[4 \cdot (\theta + 60°)] \tag{2.33}$$

where the parameters A and B are constants calculated from the maximum and minimum values of the generator inductance, corresponding to the aligned and misaligned positions of the poles, respectively.

$$A = \frac{L_{MAX} + L_{MIN}}{2} \tag{2.34}$$

and,

$$B = \frac{L_{MAX} - L_{MIN}}{2} \tag{2.35}$$

This allows for the calculation of the derivatives:

$$\frac{dL_A(\theta)}{d\theta} = -4 \cdot B \cdot sen(4 \cdot \theta) \tag{2.36}$$

$$\frac{dL_b(\theta)}{d\theta} = -4 \cdot B \cdot sen[4 \cdot (\theta - 60°)] \tag{2.37}$$

$$\frac{dL_C(\theta)}{d\theta} = -4 \cdot B \cdot sen[4 \cdot (\theta + 60°)] \tag{2.38}$$

Thus, the *Calculation* (2.30) is responsible for providing the inductance values and their respective derivatives, providing the expression (2.30) with the necessary data for the operations in expressions (2.31), (2.32), (2.33), (2.36), (2.37), and (2.38). In the *Calculation* (2.30) block, the values of currents and their derivatives, as well as the instantaneous speed and angular position, are determined and the inductance is received, as illustrated in Fig. 2.21.

For an accurate simulation, the real inductance model should be considered, as it exhibits behavior different from the sinusoidal model. Additionally, the saturation of the machine's magnetic circuit must be taken into account. To obtain these values, the following methods can be employed: (i) using the finite element method to determine $\lambda(\theta, i)$, thus creating a matrix of flux linkage values for each current amplitude and angular position; and (ii) experimentally determining the points $L(\theta, i)$ to capture the inductance behavior of the coils for various excitation current values, while also considering the saturation of the magnetic material.

This method can also employ interpolation techniques, such as Fourier series [32], to achieve more accurate results, or (iii) as an alternative, expressions

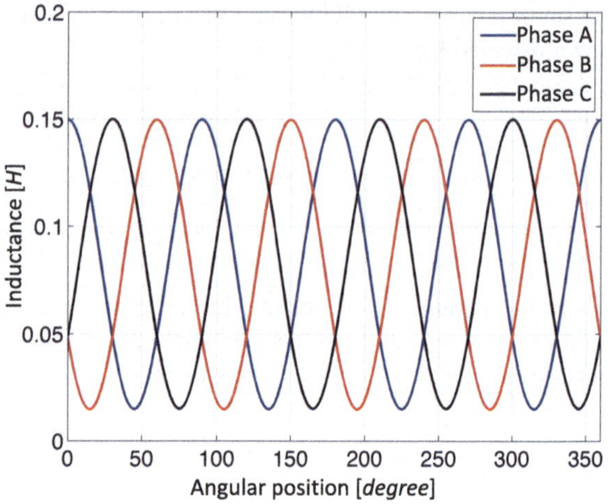

Fig. 2.21 Sinusoidal behavior of inductance per phase—linear model

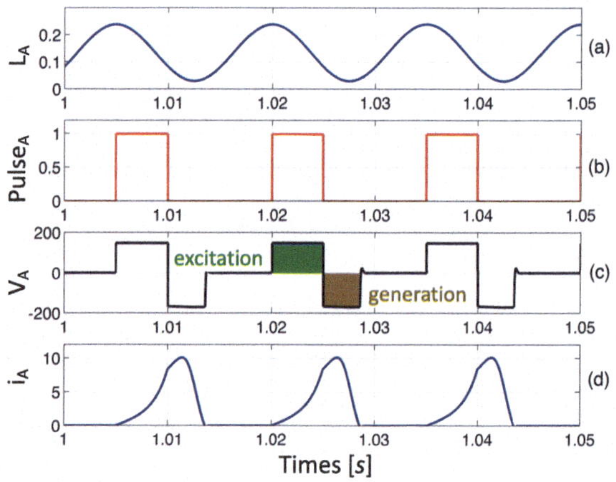

Fig. 2.22 Simulation with linear model: (**a**) inductance of phase A, (**b**) command pulses for the phase A IGBTs, (**c**) phase A winding voltage, and (**d**) current in phase A

representing the inductance surface and its respective variation rates can be derived from experimentally collected data. Although the linear model does not offer the precision of other techniques, it provides a reasonable reference for the machine's operation. Figure 2.22 shows the simulation results of the linear SRG model using the diagram in Fig. 2.20.

To enhance the simulation results obtained with the linear model, incorporating the saturation of the machine's magnetic core and accurately representing the inductance to align with the experimental model, it is proposed to represent the machine's inductance using multiple interpolated line segments. Given that the inductance of each coil is periodic, as shown in Fig. 2.22, it can be represented using a Fourier trigonometric series. Figure 2.23 illustrates how the inductance of one of the machine coils is represented by a finite number of line segments.

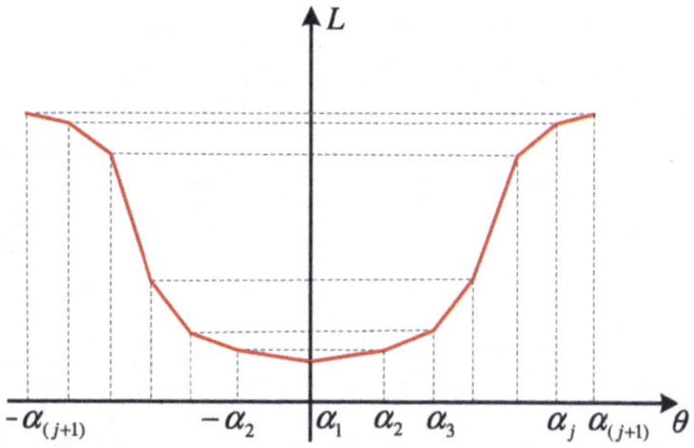

Fig. 2.23 Representation of the coil inductance of one of the SRG phases considering several straight segments

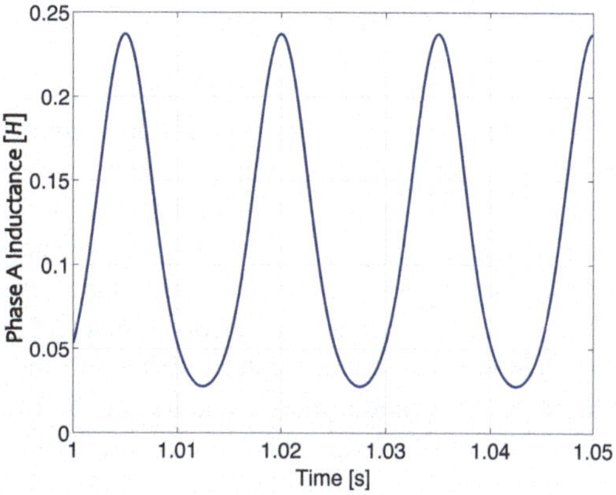

Fig. 2.24 Dynamic behavior of coil inductance for the Fourier series representation model with interpolation

In this case, the instantaneous value of the inductance is represented by:

$$L(\theta) = L_0 + \sum_{n=1}^{\infty} L_n \cdot \cos(n \cdot N_r \cdot \theta) \tag{2.39}$$

where L_0 is the constant term, and L_n is the coefficient associated with harmonic order n. N_r represents the number of rotor poles. The constant term L_0 is determined by:

$$L_0 = \frac{1}{N_r} \cdot \int_{-C_R/2}^{C_R/2} L(\theta) \cdot d\theta \tag{2.40}$$

where C_R is the pitch of the rotor pole, as defined by (2.2). The component L_n is determined by the following:

$$L_n = \frac{2}{N_r} \cdot \int_{-C_R/2}^{C_R/2} L(\theta) \cdot \cos(n \cdot N_r \cdot \theta) \cdot d\theta \tag{2.41}$$

Figure 2.24 shows the inductance behavior of one of the machine's coils as a result of the simulation using the nonlinear model described by the Fourier series.

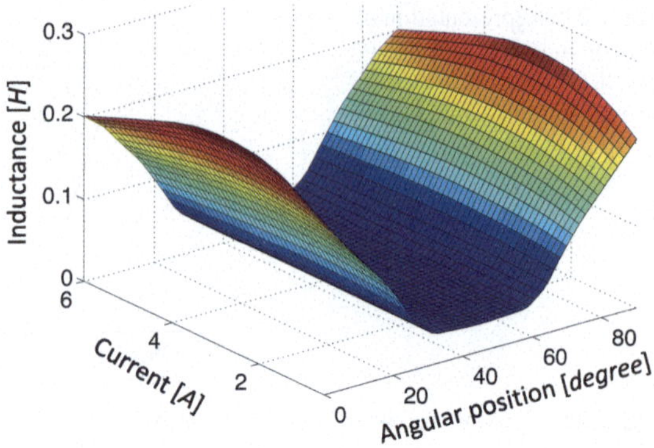

Fig. 2.25 Hypothetical illustration of the inductance surface with a set of experimentally surveyed points

Analyzing Fig. 2.24, it can be observed that the computational model based on the Fourier series, which represents the inductance of the SRG coil using a finite number of line segments, produces a distinct inductance behavior from the sinusoidal model. This observation is supported by experimental inductance measurements obtained using an impedance analyzer, as described by [reference], which reveal inductance waveforms that are more rounded at minimum values and sharper at maximum values, reflecting the construction characteristics of the generator. This model also accounts for the magnetic saturation of the machine's core, as multiple inductance curves are determined for each current amplitude, forming the inductance surface shown in Fig. 2.25.

As shown in Fig. 2.20, the computational model receives the rotor's angular position and the instantaneous current value in the coil to construct the inductance curve based on the current amplitude, reflecting the dependence between these quantities as expressed in (2.14). Numerous studies have explored computational models of SRM [4, 27]. These models share common characteristics, performing calculations using the state matrix described by (2.13). The primary differences lie in how the inductance of the SRM coils and their derivatives are represented in the simulation, which impacts the behavior of the machine's electrical and mechanical parameters. Figure 2.20 illustrates the SRM simulation flowchart based on the mathematical model in (2.13).

The SRM computational model connects the power converter circuit to the solution of the mathematical model. The comparison between the SRM model and the experimental system is fine-tuned in the **inductance profile** block, shown on the right in Fig. 2.20. The representation of the SRM coil inductance profile and its derivatives directly impacts the simulation accuracy. The simplest representation is the linear model, which does not account for magnetic saturation of the machine's core. In this model, the inductance is assumed to have sinusoidal behavior as a function of the rotor's angular position. The inductance in this case is given by:

$$L_{\varsigma}(\theta) = \frac{L_{\max_{\varsigma}} + L_{\min_{\varsigma}}}{2} + \frac{L_{\max_{\varsigma}} - L_{\min_{\varsigma}}}{2} \cdot \cos(4 \cdot \theta + \Theta_{\varsigma}) \tag{2.42}$$

where Θ [°] is the angle of displacement of the stator coils ς relative to the reference phase, and $L_{\max}[H]$ and $L_{\min}[H]$ are the maximum and minimum inductances of the SRM phase coils, respectively. In this model, the rate of state change can be easily determined. Although this model provides an understanding of the SRM's operating principle as a motor and the dynamic behavior of certain quantities, it lacks sufficient precision compared to the experimentally obtained data [2].

Nonlinear inductance models and their derivatives account for the saturation of the machine's magnetic core material [33]. To achieve this, inductance values (or concatenated flux) are calculated or measured as functions of the rotor's angular position for different excitation currents, thus defining the inductance profile. These points can be determined experimentally or using numerical methods, such as the finite element method. A widely used model in various studies is the approximation of inductance by a Fourier series, since coil inductances can be represented by periodic functions. This model approximates inductance values using multiple line segments to represent inductive behavior [33].

In the Fourier series approximation model, the positions where the inductance exhibits positive and negative derivatives do not behave as in the linear model, where the inductance follows a sinusoidal pattern. An alternative to the Fourier series model is the interpolated model, in which polynomial interpolation is applied to each line segment [27, 30]. This approach aims to smooth the inductance curve, providing improved simulation results. However, the use of interpolation increases the simulation time as a result of the additional computational effort required. All nonlinear models that account for magnetic saturation depend on experimentally acquired inductance values for various angular positions and applied current levels. These models must represent the inductance profile in the SRM simulation, as shown in Fig. 2.20 in the inductance profile block.

2.8 Control Systems for the Switched Reluctance Motor

The SRM drive has parameters that directly influence the speed and torque produced by the motor. The main parameters are: (i) excitation voltage V_{exc}, and ii)the switching angles of the coils, θ_{on} and θ_{off} [29]. In automotive applications, the SRM requires a robust control system that quickly and accurately adjusts the speed and torque produced by the motor [4]. Figure 2.26 illustrates the speed control system, which regulates the excitation voltage V_{exc} with fixed switching angles θ_{on} and θ_{off}.

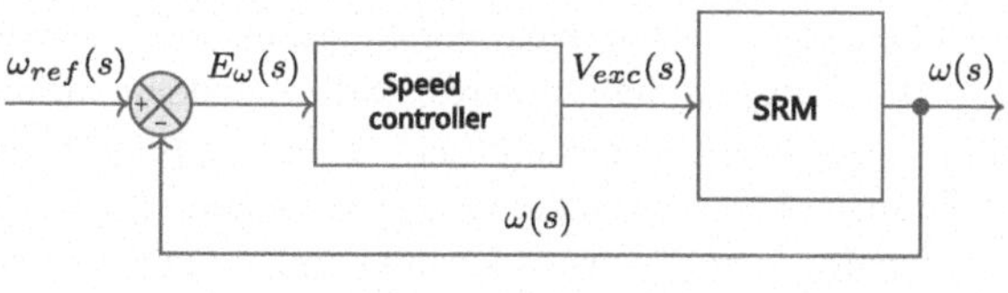

Fig. 2.26 Switched reluctance motor speed control with fixed θ_{on} and θ_{off} switching angles

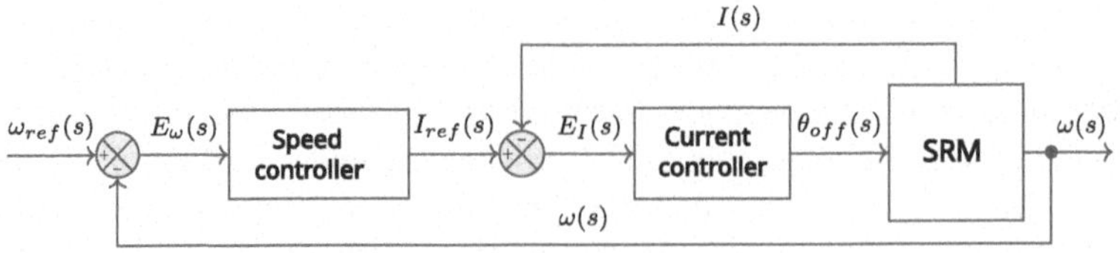

Fig. 2.27 Speed and current control of the switched reluctance motor with fixed excitation voltage V_{exc} and coil actuation angle θ_{on}

where ω_{ref}[rad/s] is the reference speed from the controller and E_ω[rad/s] is the speed error. Figure 2.27 illustrates the control system, which consists of two loops: (i) speed ω(s) and (ii) current $I(s)$. These loops control the coil turnoff angle θ_{off}, with the excitation voltage V_{exc} and the coil activation angle θ_{on} fixed [34].
where I $[A]$ is the current in the coil, E_I $[A]$ is the current error, and I_{ref} $[A]$ is the current reference. The SRM speed is compared to the reference speed, and the error is derived from this comparison. The speed controller generates the reference current, which is then compared with the current of the SRM coil, directly related to the torque produced. The power circuit operates according to the current error and the switching logic applied to θ_{on} and θ_{off}. Various techniques are used for speed control, the most commonly used being the proportional-integral-derivative (PID) controller or the *Fuzzy* logic controller (CLF). A hysteresis controller is used to control the current based on switching the coil turnoff angles [26].

The tuning of the PID controller is typically performed in the field, allowing the application of various methods to optimize the control response. Some techniques for optimizing controller tuning include: (i) deterministic or heuristic algorithms, (ii) neural networks, (iii) fuzzy logic, (iv) parametric regression, and (v) others [29]. The control method does not require prior knowledge of the mathematical model for non-optimized use, making it versatile, particularly in processes where mathematical modeling is not feasible. The PID controller is described in (2.43) for continuous time and in (2.44) for discrete time.

$$u(t) = k_p \cdot e(t) + k_i \cdot \int e(t) \cdot dt + k_d \cdot \frac{de(t)}{dt} \tag{2.43}$$

$$u(k) = k_p \cdot e(k) + k_i \cdot T_s \cdot \sum_{j=0}^{n} e(j) + k_d \cdot \frac{\delta e(k)}{T_s} \tag{2.44}$$

where e is the relative error, k_p is the proportional gain that adjusts the controller output energy, k_i is the integral gain responsible for reducing steady-state error, and k_d is the derivative gain that responds to sudden variations. In the case of discrete time, T_s represents the controller's sampling time. Hysteresis control is defined by a tolerance band around the reference current $[I_{ref} - \Delta_a, I_{ref} + \Delta_a]$, where Δ_a is the variation of the hysteresis band relative to the reference.

The reference current I_{ref} is compared with the limits of the tolerance band. If the reference current exceeds the upper limit of the band ($I_{ref} + \Delta_a$), the converter is switched to decrease the injected current. If the reference current is below the lower limit of the band ($I_{ref} - \Delta_a$), the converter is switched to increase the injected current. The main advantages of the hysteresis controller are its simplicity and robustness [4, 35].

2.9 Modeling of Fully Controlled Three-Phase Rectifier

Some studies use parametric regression to derive continuous analytical solutions within systems of numerical solutions. Experimental data collected from a test bench are analyzed, referring to a fully controlled three-phase rectifier (FCTR) with RL load, operating in discontinuous conduction. The FCTR provides a variable direct current (DC) voltage v_o from the fixed root mean square alternating current (AC) voltage at the input of the rectifier, based on the conduction angle α of the electronic switches of the converter. The solution that relates the average output voltage v_o to the conduction angle of the FCTR switches is given by:

$$v_o(\alpha, \beta) = \frac{3}{\pi} \cdot \int_{\frac{\pi}{3}+\alpha}^{\beta} V_{\max} \cdot \sin(\omega t) \cdot d\omega t \quad (2.45)$$

where $V_{\max}$ is the peak value of the phase RMS voltage of the AC input, α is the conduction angle of the switches, and β is the current extinction angle for discontinuous conduction. The current extinction angle is determined using numerical methods. To obtain the analytical solution representing the FCTR behavior, the regression method is applied, based on the following:

$$v_o(\alpha) = A_0 + \sum_{i=1}^{\kappa} A_i \cdot \alpha^{B_i} \cdot \cos(C_i \cdot \alpha + D_i) \quad (2.46)$$

where A_0, A_i, B_i, C_i, and D_i are adjustment coefficients of the regression expression, with $A_0, A_i, B_i, C_i, D_i \in \mathbb{R}$, and κ represents the number of experimental data points. The cosine expression accounts for all types of frequency and curve displacement through the coefficients C_i and D_i. This model provides a continuous analytical expression that replaces the numerical solution presented previously in the literature [36, 37].

References

1. M.R.C. Reis et al., Increasing efficiency of the switched reluctance generator using parametric regression and optimization methods, in *2017 18th International Scientific Conference on Electric Power Engineering (EPE)* (2017), pp. 1–6. https://doi.org/10.1109/EPE.2017.7967297
2. M.R.C. Reis et al., Nonlinear simulation methodology for switched reluctance machine using induction profile found by parametric regression, in *2017 CHILEAN Conference on Electrical, Electronics Engineering, Information and Communication Technologies (CHILECON)* (2017)

3. M.R. Cunha Reis et al., Optimized techniques for driving and control of the switched reluctance motor to improve efficiency. Control Eng. Practice **90**, 1–18 (2019)
4. W.R.H. Araujo et al., Efficiency enhancement of switched reluctance generator employing optimized control associated with tracking technique. Energies **14**(24), 8388 (2021) https://doi.org/10.3390/en14248388
5. J.B. Bartolo C. Gerada, The electromagnetic design of a high speed, 45kW, switched reluctance machine having a novel rotor geometry for aerospace application, in *International Conference on Electrical Machines* (2014). https://doi.org/10.1109/ICELMACH.2014.6960540
6. A. Radun, Generating with the switched reluctance motor. in *Proceedings of 1994 IEEE Applied Power Electronics Conference and Exposition - ASPEC'94* (1994). https://doi.org/10.1109/APEC.1994.316421
7. R. Krishnan et al., High reliability SRM drive system for aerospace applications, in *IECON'03. 29th Annual Conference of the IEEE Industrial Electronics Society (IEEE Cat. No.03CH37468)* (2003). https://doi.org/10.1109/IECON.2003.1280202
8. R. Cardenas, W.F. Ray, G.M. Asher, Switched reluctance generators for wind energy applications, in *Power Electronics Specialists Conference, 1995. PESC '95 Record., 26th Annual IEEE*, vol. 1 (1995), 559–564. https://doi.org/10.1109/PESC.1995.474864
9. Y. Gao, M. Mcculloch, A review of high power density switched reluctance machines suitable for automotive applications, in *2012 XXth International Conference on Electrical Machines* (2012). https://doi.org/10.1109/ICELMACH.2012.6350253
10. S.R. MacMinn, J.W. Sember, Control of a switched-reluctance aircraft en- gine starter-generator over a very wide speed range, in *Proceedings of the 24th Intersociety Energy Conversion Engineering Conference* (IEEE, 1989), pp. 631–638. https://doi.org/10.1109/iecec.1989.74532
11. E. Richter, C. Ferreira, Performance evaluation of a 250 kW switched reluctance starter generator, in *IAS '95. Conference Record of the 1995 IEEE Industry Applications Conference Thirtieth IAS Annual Meeting*, vol. 1 (IEEE, 1995), pp. 434–440. https://doi.org/10.1109/ias.1995.530332
12. J.A. Weimer, Electrical power technology for the more electric aircraft, in *[1993 Proceedings] AIAA/IEEE Digital Avionics Systems Conference* (1993), pp. 445–450. https://doi.org/10.1109/DASC.1993.283509
13. A.V. Radun, C.A. Ferreira, E. Richter, Two-channel switched reluctance starter/generator results. IEEE Trans. Ind. Appl. **34**(5), 1026–1034 (1998). ISSN: 0093-9994. https://doi.org/10.1109/28.720442
14. A. Fleury et al., Switched reluctance generator for complementary wind power generation in grid connection, in *2007 IEEE International Electric Machines Drives Conference*, vol. 1 (2007), pp. 465–470
15. R. Cardenas et al., Control of a switched reluctance generator for variable- speed wind energy applications. IEEE Trans. Energy Convers. **20**(4), 781–791 (2005). https://doi.org/10.1109/TEC.2005.853733
16. C.R. Bratiloveanu, D. Anghelus, I. Boldea, A comparative investigation of three PM-less MW power range wind generator topologies, in *2012 13th International Conference on Optimization of Electrical and Electronic Equipment (OPTIM)* (2012), pp. 535–543. https://doi.org/10.1109/OPTIM.2012.6231756
17. K. Ogawa, N. Yamamura, M. Ishda, Study for small size wind power generating system using switched reluctance generator, in *2006 IEEE International Conference on Industrial Technology* (2006), pp. 1510–1515. https://doi.org/10.1109/ICIT.2006.372468
18. M. Nassereddine, Switched reluctance generator for wind power applications. Master of Engineering. School of Engineering - University of Western Sydney, 2011
19. H. Chen, Z. Shao, Turn-on angle control for switched reluctance wind power generator system, in *30th Annual Conference of IEEE Industrial Electronics Society, 2004. IECON 2004*, vol. 3 (2004), pp. 2367–2370. https://doi.org/10.1109/IECON.2004.1432170

20. M.A. Mueller, Design and performance of a 20 kW, 100 rpm, switched reluctance generator for a direct drive wind energy converter, in *IEEE International Conference on Electric Machines and Drives, 2005* (2005), pp. 56–63. https://doi.org/10.1109/IEMDC.2005.195701
21. T.J.E. Miller, Optimal design of switched reluctance motors. IEEE Trans. Ind. Electron. **49**(1), 15–27 (2002). ISSN: 0278-0046. https://doi.org/10.1109/41.982244
22. R. Krishnan, *Switched Reluctance Motor Drives: Modeling, Simulation, Analysis, Design, and Applications* (CRC Press, Boca Raton, 2017)
23. D.A. Andrade et al., Energy efficiency for fractional power loads. IEEE Ind. Appl. Mag. **12**(6), 12–20 (2006). ISSN: 1077-2618. https://doi.org/10.1109/IA-M.2006.248009
24. A.M.S. Arifin, Switched reluctance generator drive in the low and medium speed operation: modelling and analysis. Thesis. Massey University, 2013
25. C. Pollock, A. Michaelides, Switched reluctance drives: a comparative evaluation. Power Eng. J. **9**(6), 257–266 (1995). ISSN: 0950-3366. https://doi.org/10.1049/pe:19950606
26. L.D.S. Morais et al., Enhancement of the performance of switched reluctance generators in low wind speed conditions using advanced tracking techniques. Int. J. Energy Res. **2024**(1), 5541150 (2024)
27. R. Krishnan. *Eletric Motor Drives: Modeling, Analysis, and Control*, vol 1 (New Jersey, 2001)
28. G. Chen, T.T. Pham, *Introduction to Fuzzy Sets, Fuzzy Logic and Fuzzy Control Systems*, vol. 1 (CRC Press, Boca Raton, 2000)
29. M.R.C. Reis et al., Analysis of switched reluctance motor efficiency under different speed control strategies, in *2015 CHILEAN Conference on Electrical, Electronics Engineering, Information and Communication Technologies (CHILECON)* (IEEE. 2015), pp. 467–472
30. Z. Wang et al., Electromagnetic modeling and investigation for bearingless switched reluctance motor considering magnetic saturation. IEEE Trans. Energy Conversion **38**(1), 122–133 (2022)
31. N. Schofield et al., Design of a switched reluctance machine for extended speed operation. IEEE Trans. Ind. Appl. **45**(1), 116–122 (2009)
32. D.A. Andrade, R. Krishnan, Characterization of switched reluctance ma chines using Fourier series approach, in *Conference Record of the 2001 IEEE Industry Applications Conference. 36th IAS Annual Meeting (Cat. No.01CH37248)*, vol. 1. (2001), pp. 48–54. https://doi.org/10.1109/IAS.2001.955391
33. D.A. Andrade, R. Krishnan, Characterization of switched reluctance machines using Fourier series approach, in *Conference Record of the 2001 IEEE Industry Applications Conference. 36th IAS Annual Meeting (Cat. No. 01CH37248)*, vol. 1 (IEEE. 2001), pp. 48–54
34. N.S. Nise, *Control Systems Engineering* (John Wiley & Sons, New York, 2020)
35. K. Ogata et al., *Modern Control Engineering* (Prentice Hall India, 2009)
36. I. Barbi, F. Pöttker, *Soft Commutation Isolated DC-DC Converters*, vol. 1 (Springer, Berlin, 2019)
37. H. Rashid Muhammad et al., *Power Electronics-Circuits, Devices, and Applications* (Pearson Prentice Hall, Upper Saddle River (2004)

Chapter 3
System, Modeling, Simulation, and Optimization

3.1 Concept of System and Model

A system is a set of interacting elements, constituting an organized entity as a whole with connections to other entities and that exists within an environment. A system is a collection of components that produce behavior or functions that are not possible by any single component individually. Systems can be classified into three domains: (i) systems in the world, (ii) systems in the mind, and (iii) systems in the abstract. These domains refer to ontological, epistemological, and mathematical/symbolic aspects, respectively. The systems that constitute reality are only abstracted mathematically and symbolically after mental representation, as illustrated in Fig. 3.1.

The mental and abstract representations of systems are known as models. The creation of models as useful maps of the world involves a process of simplification, separation, and identification. For the authors, this process is inherently uncertain [1]. Even if a model is extensively corroborated, resisting formal tests, internal consistency checks, or assessments of its explanatory or predictive capacity, it cannot be validated as an exact representation of the system as it constitutes an abstraction of reality. A single modeler may represent a given system through different models, depending on the level of abstraction applied or the selection of parts of interest. These parts generally constitute subsystems, each with its own set of elements, boundaries, and relationships. Regardless of the hierarchical level in question, systems exhibit common patterns and behaviors that are studied within systems science.

W. P. Calixto et al., *Control and Tracking Techniques for Switched Reluctance Machines*, Power Systems, https://doi.org/10.1007/978-3-031-86727-9_3

Fig. 3.1 Domains of systems

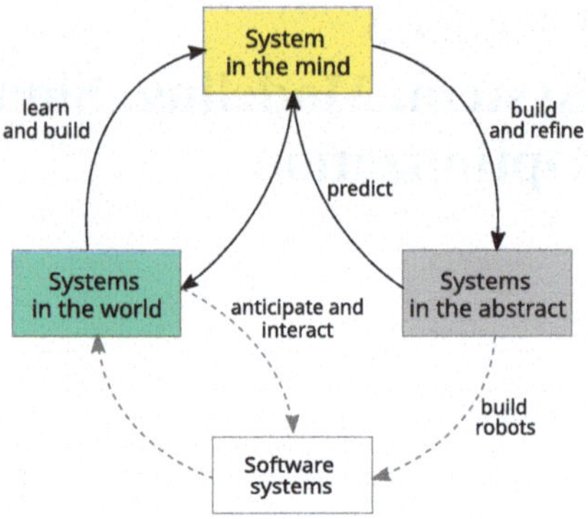

3.2 Systems Science

Given the importance of the term system, which is widely used, a field of scientific inquiry was established with systems as its main object of study, known as Systems Science [2]. This field seeks to understand systems as they are, identifying continuities across differences (linear, nonlinear, closed, open, complex) rather than segregating systems into groups as is commonly done [3]. Systems science is classified as a formal science due to its mathematical conceptions of systems. However, its manifestations (e.g., physical, biological, cognitive systems) are framed within phenomenological or factual sciences and relate directly to certain disciplines, as illustrated in Fig. 3.2 [4]. Although systems science encompasses theory on systems, including mathematical definitions, the field of systems design

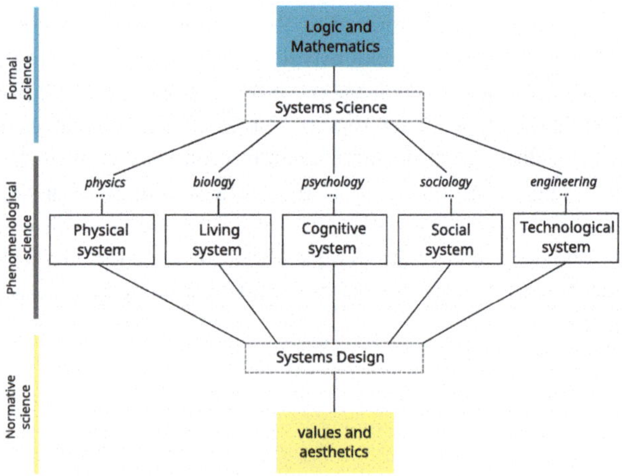

Fig. 3.2 Classification of sciences and systems

covers general methods and practical applications (in human-created systems), thus being classified as a normative science [4].

Through the investigation of general attributes, dynamics, characteristics, and behaviors of systems, Systems Science proposes principles that govern all systems. Among these principles is the term systemness, used in sociology [3]. According to the authors, systemness is a recursive property in which a system can simultaneously encompass smaller systems and be encompassed by larger systems, yet each system retains its own wholeness. The principles of Systems Science state that systems: (i) possess systemness, (ii) are organized processes within structural and functional hierarchies, (iii) can be represented as networks of relations among components, (iv) are dynamic across multiple time scales, (v) exhibit various types and levels of complexity, (vi) evolve, (vii) encode knowledge and share information, (viii) contain regulatory subsystems to achieve stability, (ix) include models of other systems, (x) if adaptive, may contain models of themselves, (xi) can be understood (Science), and (xii) can be improved (Engineering).

Based on these principles, various subfields within systems science have been developed, including classical thermodynamics, open systems theory, information theory, cybernetics, autopoiesis theory (self-production), chaos theory, multiagent modeling and autonomous agents, complexity theory, and network science [4]. Despite their specificities, systems exhibit common properties related to their composition and organization, forming a unified whole.

3.3 Properties of Systems

Given that a system can encompass or be encompassed by particular systems, what defines it as a system in itself is the combination of basic properties. In general, systems possess a bounded and organized structure that establishes internal and external relationships. These properties are universal and include: wholeness, composition, internal organization, and external organization [3].

3.3.1 Wholeness

Wholeness refers to the internal cohesion among elements bounded by a border that may be concrete, porous, diffuse, or conceptual. To determine what enters and exits the system, boundary conditions are established that control the exchange of material, energy, and information with other systems. Any system viewed as an entity must have boundaries, which are ultimately dynamic. Due to the law of recursion (or the principle of systemness), systems overlap and are nested, forming a unified totality in which entities interact with others at the same ontological level or within the same level of recursion [3, 5, 6].

3.3.2 *Composition*

Composition refers to the elements and interactions that make up a system. The elements, or components, are generally subsystems viewed as entities with potential interactions defined by their boundary conditions and may be treated as black boxes (with unknown internal structure). Each type of component has its own uniqueness, known as a personality, which expresses its specific capacity to interact with other types of components. Thus, the greater the heterogeneity of components within a system, the higher the complexity of its internal structure and function. In addition, the same component can serve multiple purposes at different times, allowing functional overlap as a safety mechanism for critical functions [3].

3.3.3 *Internal Organization*

The internal organization of systems refers to how the structure performs functions or processes. Although structure refers to patterns that persist over time, organization refers to the dynamics established internally. In systems, elements are organized through patterns of connections or networks, represented by vertices (elements) and edges (connections). Thus, the connection linking two elements has a specific coupling strength,[1] determined by the impact a change in one component has on the other [3, 7]. Components can be connected in various ways, such as association, juxtaposition, concatenation, grouping, or addition. Among these forms, association is the most common in generating combinatorial novelty and plays a crucial role in most complex systems. In addition to connectivity among elements, internal organization is based on hierarchy, as components may be subsystems, which in turn may consist of subsystems, and so on, until partitioning ceases and elementary subsystems are defined. Together with these characteristics, or as a consequence thereof, complexity manifests itself in systems at various forms and levels [8, 9].

3.3.4 *External Organization*

Systems interact in multiple ways, establishing upward, downward, lateral, and external connections with similar or different systems that may or may not belong to the same unified whole. When a system is viewed as a whole within a larger system, some of its characteristics become input or output parameters in the system–environment relationship, mediated by boundary conditions. Throughout the system lifecycle, the environment provides resources (input) and receives waste (output).

---

[1] The connection between any two components of the system.

Resources refer to everything necessary for the system's operation (matter, energy, or information), while waste consists of the by-products of processed resources. Thus, environmental flows contribute to understanding the behavior of the system while reflecting its internal organization [3, 6].

3.4 Modeling and Simulation

Investigations into system behavior rely on observation or, more specifically, on experiments. In both cases, models are commonly used to abstract essential aspects of a system's organization and its relationship with the environment. With a model in hand, it is possible to simulate the functioning of the system of interest under typical or rare conditions by manipulating variables and observing the resulting implications. This approach enables the understanding of the behavior of the system under specific conditions, leading to solutions to problems whose causes were previously unknown or poorly understood [1, 10, 11]. Natural or artificial systems are governed by rules that direct their behavior. Modeling is, therefore, the process of representing these rules through a set of structures, called a model. The model is a simplified representation of the various interactions between the components of the system, capturing only what is truly important for the intended purpose [1, 10].

The model can be associated with experiments and implemented as a computer program, also known as a simulation model or simulator. The accuracy of the simulator is verified through tests that assess whether its behavior aligns with the model specifications in as many cases as possible. Similarly, the model is validated by comparing the simulator's results with those produced by the real system. Thus, both the simulator and the model can be adequately evaluated using verification and validation techniques, respectively [11].

The following steps comprise the modeling and simulation process: (i) problem formulation, (ii) creation of the conceptual model, (iii) input and output data collection and analysis, (iv) modeling, (v) simulation, (vi) verification and validation, (vii) experimentation, and (viii) output analysis. Problem formulation involves contextualizing the system of interest, identifying the experimental framework and input/output variables, as well as defining performance metrics and the objective function when performance analysis is required. The experimental framework encompasses the experimental conditions, parameter values, and mechanisms that generate the behavior of the system [11].

Once the problem has been formulated, the conceptual model is created, consisting of a description of the system's structure and behavior, identifying all elements along with their attributes and interfaces. In this phase, state variables, critical characteristics, and non-functional information are also defined. In the subsequent phase, input and output data collection and analysis, the data sample size, and attribute types (stochastic or deterministic) are determined prior to data collection and analysis. When no data source is available, similar systems may be used to obtain datasets, or a stochastic approach may be employed to generate them. These data, along with the experimental framework and the conceptual model, form the foundation for modeling, which is the process of developing a detailed representation of the system [11].

The next phase is simulation, where the model is implemented to create the simulator, combined with the model of the experimental framework. After verifying the simulator and the validation of the model, the experimentation process can begin with experiment planning, using techniques such as sensitivity analysis, optimization, variance reduction, classification, and selection. In this phase, the objective is to run the simulator as many times as necessary to conduct the experiment. Once this phase is complete, the output analysis follows, with the aim of understanding how the output values provide insight into the behavior of the original system.

During experimentation, optimization methods can be applied to solve problems in the context of systems. Such methods may be associated with classes of systems (simple, complicated, complex, and chaotic) as represented in the Stacey matrix in Fig. 3.3. Complex systems can be optimized using heuristic, inferential, and stochastic methods, as they fall within regions characterized by disagreement (regarding the system itself and the ways to describe it) and uncertainty (in its dynamics). This uncertainty also pertains to parameter variations due to changes in environmental conditions. The effects of these variations on the output of the system are measured through sensitivity analysis [12, 13].

The complexity of systems is associated with their dynamic and stochastic nature, meaning that their state changes over time or through adaptation, as well as the presence of random variables in their design. Thus, a simulation model can reproduce these characteristics by exhibiting behavior consistent with the original system when subjected to the same boundary conditions. Simulation models can be computationally classified into three basic categories: Monte Carlo simulation, continuous simulation, and discrete-event simulation. Monte Carlo simulation relies on generating random numbers for the variable values, but time is not explicitly considered as a variable. In contrast, simulations of continuous and discrete events

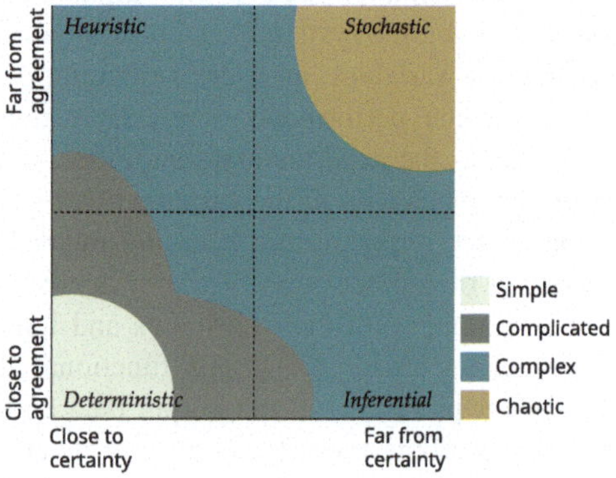

Fig. 3.3 The Stacey matrix and its relation to decision support methods

Table 3.1 Classification of models based on time and state variable foundations

Variable/Time	Time continuous	Time discrete
Variable continuous	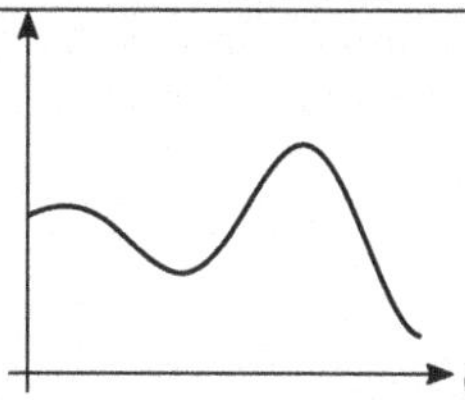 (a)	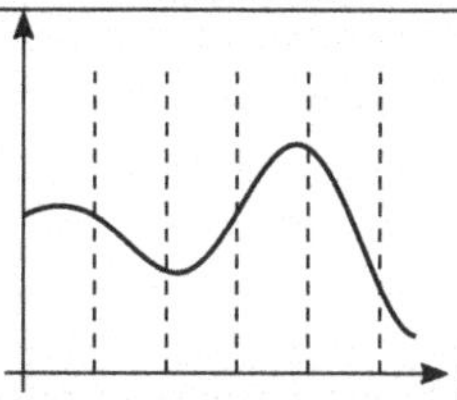 (b)
Variable discrete	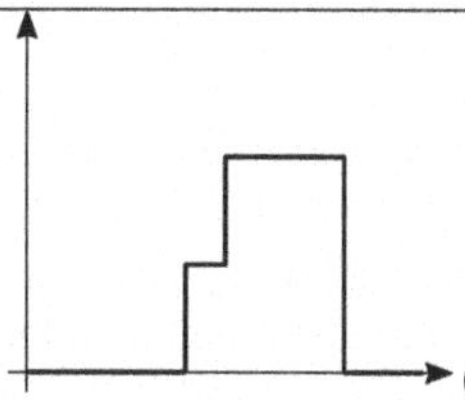 (c)	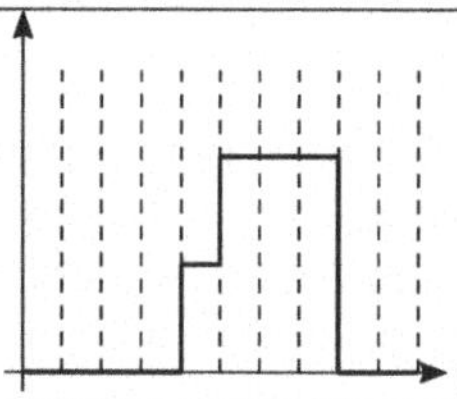 (d)

involve continuous and discrete variables, respectively, which are observed over time [3, 10, 14].

Different techniques of discrete or continuous modeling and simulation are categorized on the basis of the representation of state variables and time within the model. When time is the basis, techniques consider that time evolves either continuously (continuous time) or in discrete intervals (discrete time). For state variable values, variables can be represented by a continuous set of values (continuous variables) or by a finite set of values (discrete variables), depending on the model. Table 3.1 graphically presents these cases [10].

There are modeling techniques appropriate for each classification in Table 3.1. Commonly used techniques include: for the cases in Table 3.1(a), partial or ordinary differential equations, bond graphs, and electrical circuit diagrams; for the cases in Table 3.1(b), difference equations, finite element methods, finite differences, and numerical methods; for the cases in Table 3.1(c), discrete event system formalisms, timed Petri nets, timed finite state machines, and event graphs are employed; and for the cases in Table 3.1(d), finite state machines, finite state automata, Petri nets, Boolean logic, and Markov chains are used [11].

In systems with discrete variables, state transitions are driven by events. In models of such systems, discrete variables represent states and are defined within qualitative domains. Discrete event systems with a continuous time basis are dynamic, concurrent, and highly nonlinear, typically modeling human-made creations, such as traffic controllers, robotic arms, automated factories, production plants, and computer networks, among others. In cases where time is also discrete, events occur at fixed intervals [11, 15].

Time can also be an implicit variable in simulation through time series, as exemplified in Monte Carlo methods. This type of simulation is independent of the system's evolution over time, as it evaluates system outputs for different sets of randomly generated input according to probabilistic distributions representing the variables in question. Probabilistic models are used in critical situations where deterministic models are inefficient, such as in the simulation of physical systems with a high number of degrees of freedom, complex dynamics, and specific time scales. For this reason, Monte Carlo simulation has been widely applied to both physical systems and models [16].

3.5 Applied Optimization Process

The essence of the system optimization process is to find the best solution among the various possible solutions available. Facile solutions are those that satisfy all the constraints of the optimization problem. The best solution might involve minimizing the cost of a given process or maximizing the efficiency of a certain system. Current optimization challenges are multidisciplinary and multiobjective. Optimization techniques used to solve these problems include not only gradient-based methods but also natural-inspired processes, such as genetic algorithms, ant colonies, and particle swarms, among others [17]. In optimization problems, a function must be maximized or minimized. Typically, only minimization is used, as a maximization problem can be transformed into a minimization problem. For a maximization problem of $f(x)$, the optimal value to achieve is ∞. Thus, if $f(x)$ is a function defined in $\mathbb{R}$, we have:

$$\lim_{f(x)\to 0} \frac{1}{\min f(x)} = \infty \tag{3.1}$$

where $\min f(x)$ represents the minimization problem. This approach reformulates a maximization problem as a minimization problem. To define the minimization problem, where the optimal point will always be zero, it suffices to calculate the modulus $\mid f(x) \mid$, ensuring that whether the goal is maximization or minimization, zero will always be the global optimal point [18]. This process simplifies matters for the developer, as in most cases the exact value of $f(x)$ to be achieved in the optimization process is unknown. The function $f(x)$ being optimized is called the objective function or the evaluation function. The values assumed by x in $f(x)$ are referred to as design variables or decision variables. Practical considerations require these variables to remain within a specific range, bounded by a lower limit x_L and an upper limit x_U. The optimization problem can be mathematically expressed as:

$$\begin{gathered} \min f(x) \\ \texttt{sujeito}: \\ g_i(x) \leq 0, \ i = 1, 2, \cdots, m \leq n \\ h_j(x) = 0, \ j = 1, 2, \cdots, r \leq n \\ \mathtt{com}\, x \in \mathbb{R}^n \subset \Omega \end{gathered} \tag{3.2}$$

where Ω is the set in which solutions exist and are feasible, given the constraints within the search space, and x is the vector with n decision variables (parameters to be optimized), defined as:

$$x = \begin{bmatrix} x_1 \\ x_2 \\ \dots \\ x_n \end{bmatrix} \tag{3.3}$$

The functions $f(x)$, $g_i(x)$, and $h_j(x)$, in deterministic methods, must be differentiable, and x should be bounded by x_L and x_U. The constraints $g_i(x)$ are inequality constraints, while $h_j(x)$ are equality constraints. The constraints of the optimization problem are functions of the decision variables. Additionally, the number of constraints $g_i(x) + h_j(x) \leq n$. If the design variables satisfy all the constraints, they form the feasible set, and any element within this set is referred to as a feasible point [19]. If the objective function and constraints are linear functions of x, the optimization problem is classified as a linear optimization problem. In most cases, objective functions are multimodal, which means that they have multiple optimal points, both local and global. Figure 3.4 illustrates several global and local maxima and minima, showing isoline values on the $x_1 \times x_2$ plane, where the red $\times$ indicates the global maximum, and the blue $\times$ represents the global minimum.

The development of optimization algorithms is a task that requires both skill and time from the programmer. The algorithms used to solve optimization problems can be deterministic, heuristic, or stochastic. Optimization aims to make the process as efficient as possible. To perform optimization, knowledge of the process and a clear definition of the variables to optimize are necessary [18, 19]. The optimization problem can be expressed as:

$$\min\ f(x) \text{ sujeito à: } \begin{cases} c_i(x) = 0, i \in X \\ c_i(x) \neq 0, i \in Z \\ \text{com } x \in \Re^n \subset \Omega \end{cases} \tag{3.4}$$

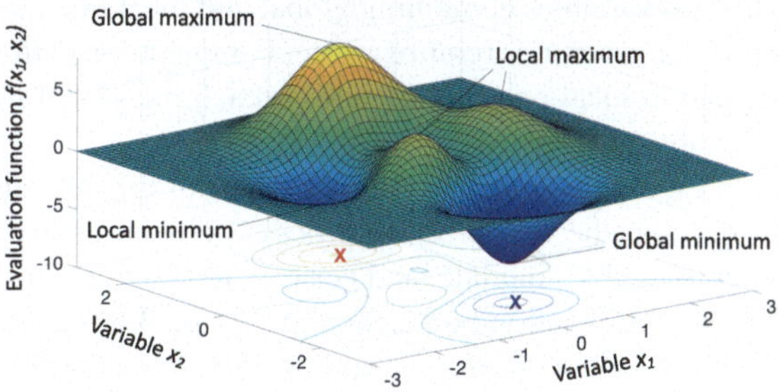

Fig. 3.4 Local and global optima for a two-dimensional function

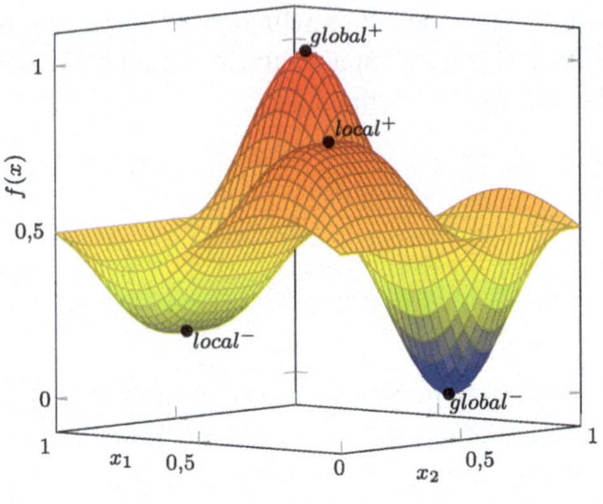

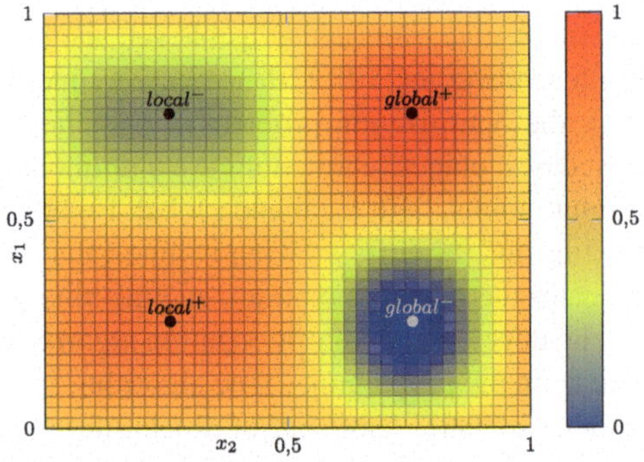

Fig. 3.5 System optimality surface with two variables: (**a**) three-dimensional view and (**b**) top view

where x is the vector of variables subject to optimization, f is the evaluation function of the process, c_i are the constraint functions, X and Z represent the sets of indices for equality and inequality constraints, respectively, n is the number of parameters to optimize, and Ω is the set in which solutions exist and are feasible given the constraints (search space) [20]. Most problems have evaluation functions with local and global optima. Figure 3.5 illustrates the optimality surface of a system with two variables containing global and local maxima and minima. Problems with these characteristics require greater computational effort, as algorithms may become trapped in local optima and fail to reach the global maximum or minimum on the optimality surface.

In Fig. 3.6, $f(x)$ represents the value of the evaluation function for the problem. To develop the optimization process, it is necessary to understand the system in order to model, simulate, and perform tests. Only after the model has been validated can the optimization be performed. Figure 3.6 illustrates the optimization process, where f^* is the optimal or optimized value of the evaluation function, and $f(x^*)$ represents the optimal or optimized parameters [10, 21].

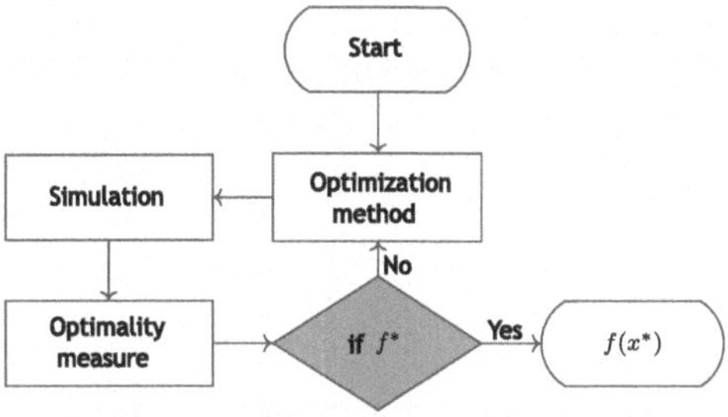

Fig. 3.6 Fluxograma do processo de otimização

3.6 Optimization Methods

Deterministic optimization methods generate a sequence of potential solutions, often requiring the use of at least the first derivative of the objective function, which must be continuous and differentiable within the search space Ω. These methods guarantee convergence to an optimal solution, though not necessarily the global optimum. They employ analytical algorithms that are based on knowledge of the derivatives of the objective function at each iteration. The objective function and the constraints are defined as mathematical functions and functional relationships. These methods are sensitive to the initial estimate, which can make them inefficient in finding global optima for some problems [22]. However, they possess theoretical guarantees of convergence to a local optimum. Examples include the Newton method, the quasi-Newton method, and the gradient method. These methods exhibit predictable behavior, providing the same output and stopping point for a given input [19, 23].

Heuristic methods are based solely on the evaluation function and introduce random data and parameters into the optimization process. No derivative information from the evaluation function is used. The advantages of heuristic methods over deterministic methods include: (i) the evaluation function and constraints do not necessarily require a mathematical representation, (ii) they do not require the evaluation function to be continuous, (iii) they work effectively with both continuous and discrete parameters, (iv) there is no restriction on the starting point within the search space, and (v) they can optimize a large number of variables. The disadvantage of heuristic methods compared to deterministic methods for linear systems is the longer processing time. However, this disadvantage may be outweighed by the difficulty that deterministic methods face in solving highly complex nonlinear problems. Heuristic methods involve continuous empirical searches that, even with multiple local optima, yield the best solution attainable under specific conditions [24, 25].

Deterministic optimization methods employ analytical techniques to minimize the evaluation function. These methods rely on knowledge of the derivatives of the evaluation function and the initial estimate (initial values of x). Such techniques are based on the gradient of the evaluation function, specifically the Jacobian or Hessian matrix [18]. This provides the search direction necessary to locate the maximum or minimum of the function. As most practical applications are nonlinear, the search for $f'(x)$ is performed using numerical techniques to approximate the derivative at a given point. If $f(x)$ is known at a point x, then the value of the function in the neighborhood $x + \Delta x$ can be calculated using the Taylor series, given by:

$$f(x + \Delta x) = f(x) + \Delta x \cdot f'(x) + \frac{\Delta x^2}{2!} \cdot f''(x) + \frac{\Delta x^3}{3!} \cdot f'''(x) + \dots \quad (3.5)$$

This applies to a function with only one variable. If more variables are involved, the gradient of $f(x)$ can be obtained through the partial derivatives with respect to each variable:

$$\nabla f = \begin{bmatrix} \frac{\partial f}{\partial x_1} \\ \frac{\partial f}{\partial x_2} \\ \dots \\ \frac{\partial f}{\partial x_n} \end{bmatrix} \quad (3.6)$$

For constrained optimization problems, a change in the direction of the gradient may lead to migration to an infeasible region. In such cases, it is desirable to move in an alternative search direction and to understand the rate of change of the function in this direction. Directional derivatives provide information on the instantaneous rate of change of a given function in a particular direction. Although these methods can be used to find the optimal solution for a given problem, they have certain drawbacks, such as: (i) difficulties in finding global optimal solutions, (ii) challenges in problem analysis when discrete variables are involved, (iii) the requirement for differentiable functions, (iv) increasing difficulty in obtaining the optimal point numerically as the number of variables grows, and (v) limited effectiveness in addressing multiobjective optimization problems.

The quasi-Newton method (QNM) is based on Newton's method, starting from initial estimate values and subsequently developing a quadratic approximation of the evaluation function. The minimization of the quadratic function is exact, and the minimum point of this quadratic function indicates a possible point leading to the lowest value of the evaluation function. This procedure repeats until the algorithm stops and the criterion is met. In QNM, the gradient of the objective function is used at each iteration. These methods are globally convergent if the step length satisfies Wolfe's condition and if the pseudo-Hessian matrices are numerically bounded and positive definite. The quasi-Newton method combines the speed of the gradient method with the accuracy of Newton's method. For applying the quasi-Newton algorithm to discrete optimization problems, the evaluation function must be constructed as continuous [22, 23].

Heuristic optimization methods are random search methods in which each iteration aims to approach the optimal or optimized solution. Like deterministic methods, they use the evaluation function but offer several advantages, such as: (i) they do not require the evaluation function to be continuous or differentiable; (ii) they work with continuous parameters, discrete parameters, or combinations of both; (iii) there is no restriction on the starting point within the search space; (iv) they can optimize a large number of variables with relative ease of computational implementation; and (v) they are flexible in handling arbitrary constraints and optimizing multiple functions. Heuristic methods are more robust than deterministic methods as they allow for optimized solutions even in problems where the optimality surface contains multiple local optima. However, they require higher computational performance and longer processing times. Among optimization techniques that employ random search methods, genetic algorithms (GAs) stand out [26, 27].

The genetic algorithm is a search optimizer based on the mechanism of natural selection. It relies on Darwin's key principle of survival of the fittest. The initial population of individuals is randomly generated and undergoes various transformations. After several generations, the goal is for the population to evolve such that the best individual represents the optimal or optimized solution. The genetic algorithm (GA) mimics the genetic process in which hereditary traits are passed from parents to offspring. The basic unit of inheritance is the gene, represented here by the values $x_1, x_2, \cdots, x_n$ within the vector x. In deterministic methods, the solution moves from one point to another using the gradient, as indicated in (3.6). The GA, on the contrary, operates on a population of points rather than a single point. The evaluation function is calculated for each individual in the population, each representing a potential solution to the problem. The best individual from each population is copied into the new population through a process known as elitism. Elitism ensures that, in the worst case, the best individual in the next generation $g + 1$ is at least as good as the best individual in the previous generation g [28].

After elitism, individuals are selected as parents. Selection is typically based on the evaluation function value of each individual. With the chosen parents, the first operation, the crossover, is performed. In this operation, genes are recombined between parents to generate a new individual in the hope of producing better offspring. Better offspring indicate that the new individuals have an improved evaluation function compared to their parents. Once the crossover operator has been applied, the new population is formed. The mutation operator is then applied to this new population. Mutation introduces a perturbation in the gene's value, modifying it up or down, depending on a binary random decision variable. The best individual is then inserted into the new population, initiating a new cycle of individual evaluation. Figure 3.7 presents the flowchart of GA functioning.

The efficiency of the genetic algorithm depends on exploring new regions of the search space and exploiting previously visited regions. The genetic algorithm begins its search once the internal parameters are defined: (i) population size, represented by an $m \times n$ matrix in which the population matrix contains m individuals with n genes; (ii) crossover rate, defined as the probability of crossover between parent data; a higher rate increases the inclusion of new individuals in the population; (iii) mutation rate, defined as the probability of mutation occurring on a given

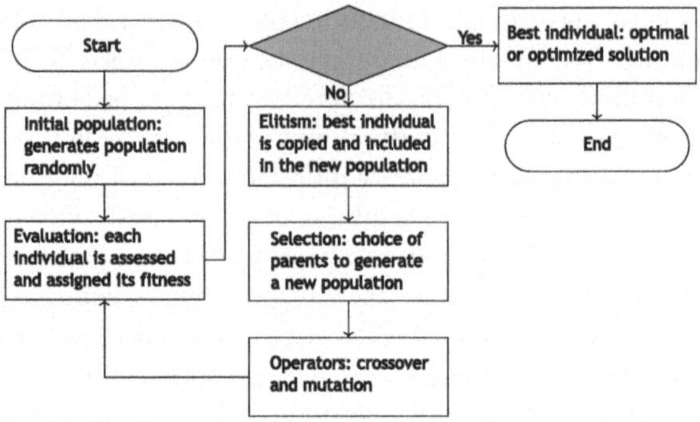

Fig. 3.7 Flowchart of the Classical Genetic Algorithm

chromosome; (iv) stopping criterion, which indicates when the search process should terminate; and (v) maximum number of generations g.

The crossover operator is applied to the population with the objective of global exploration of the search space **exploration**. The mutation operator is applied to the population to locally explore the search space **exploitation**. These processes end when the best solutions to the problem are found or when $g = g_{max}$. The genetic algorithm (GA) belongs to the group of heuristic optimization methods. Based on populations and the exchange of information between individuals, these methods are inspired by evolutionary theory and are applied to problem solving. The genetic algorithm operates with the characteristics: (i) each individual represents a potential solution to the problem, (ii) a group of individuals forms a population, and (iii) each individual is assigned a fitness score based on the evaluation function, which indicates its closeness to the optimal or optimized solution. The genetic algorithm uses selection methods to prioritize the reproduction of individuals with higher fitness. Genetic recombination is then performed in which the parent individuals combine to create a new generation [29].

In genetic algorithms, the chromosome is the data structure that represents potential solutions within the problem search space. Chromosomes (individuals in the population) undergo a process that includes evaluation, selection, recombination, and mutation. Starting the genetic algorithm with the same initial population and parameter set may yield different solutions for each execution. Genetic algorithms work with point populations, applying a search heuristic in the solution space Ω. During the execution of the genetic algorithm, the population of individuals, representing the set of potential solutions to the problem, is subjected to a series of transformations and operations. Each evaluation cycle constitutes a generation. The expectation is that, by the end of the maximum number of generations, the genetic algorithm will yield an optimal candidate or that the best individual will represent the optimized solution. The structure of the basic genetic algorithm is summarized as shown in Fig. 3.7 [18, 30].

In the classical genetic algorithm, the initial population is generated randomly. In certain specific cases, the genetic algorithm can be initialized with an evolved population or with one generated by an initialization algorithm directed toward the optimum. The evaluation function serves as the score, or fitness, assigned to each individual based on its quality in solving the problem. This evaluation function should embed all relevant knowledge about the problem to be solved. Various stopping criteria can be used, such as a predefined number of generations or a value of the target evaluation function. The elitism method ensures that the best individual is preserved, preventing it from being lost due to genetic operator manipulation [30].

Elitism ensures that the performance of the genetic algorithm improves over successive generations. The best individual from each generation should not be replaced, but rather passed on to the next generation to preserve its genetic information. Selection is the process of choosing parents. Tournament selection is one of the most refined selection methods, as it allows adjustment of the selection pressure, thereby controlling the degree to which the selection method considers the evaluation value of individuals. In tournament selection, individuals compete against randomly chosen opponents from the population, and the individual with the best evaluation function wins the competition based on the number of victories achieved [18].

The crossover operator combines the genes of two or more individuals, allowing future generations to inherit traits from individuals from previous generations. The concept behind this operator is the exchange of genetic information between individuals. The heuristic crossover operator is one method that generates a new individual containing the characteristics of the fittest parents. The distance between the newly generated individual (offspring) and the less fit parent is established, allowing the characteristics of the new individual to be defined relative to those of the parents. In genetic algorithms, the mutation operator serves the purposes: (i) to prevent the permanent loss of chromosomes, (ii) to maintain diversity throughout the algorithm's execution, and (iii) to introduce energy to escape local optima. One mutation operator found in the literature is the adaptive mutation operator, which applies random mutations to individuals. With the adaptive mutation operator, each individual's value is modified by a random variable, using a probability function with a mean of zero and an adaptable standard deviation [22].

In optimization, hybridization involves the use of more than one type of optimization algorithm within the same process. Typically, hybridization occurs through the exchange of solutions between algorithms. When a deterministic algorithm is combined with a heuristic algorithm, the process generally begins with the heuristic algorithm, which, after a predetermined number of iterations, passes its best solution to the deterministic algorithm. After completing its iterations, the deterministic algorithm returns its best solution to the heuristic algorithm. This process continues until the stopping criterion is met. In the case of the genetic algorithm and the quasi-Newton method, the process begins with the initial population generation of the genetic algorithm. After a specified number of generations $g_{\max}$, the genetic algorithm provides the quasi-Newton method with the best individual. The quasi-Newton method uses this best individual as its initial estimate, representing an already evolved solution. Upon completion, a randomly selected individual from the population of the genetic algorithm is replaced by the solution obtained by the quasi-Newton method. This hybridization process is illustrated in Fig. 3.8 [27].

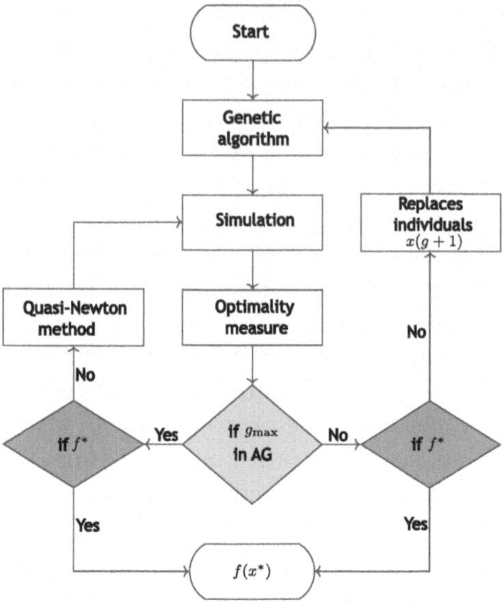

Fig. 3.8 Flowchart of the hybridization process

Hybridization is used to: i) improve the performance of existing techniques and ii) enhance the search for solutions that individual methods or algorithms may not find independently [21]. Any technique can be selected for hybridization, which means that hybridization can involve various techniques to solve different problems.

3.7 Wind Power Generation System Model

Wind energy is defined as the energy contained in moving air masses. It is harnessed through wind turbines that convert the translational kinetic energy of air masses into rotational kinetic energy [31, 32]. This rotational kinetic energy is then converted into electrical energy using electric generators. The kinetic energy E_c of the moving air masses is given by:

$$E_c = \frac{1}{2} \cdot m \cdot v_v^2 \tag{3.7}$$

where m is the mass of the air, and v_v is the speed of the wind. The mechanical power P_t extracted by the wind turbine is given by:

$$P_t = \frac{1}{2} \cdot \rho \cdot A \cdot v_v^3 \cdot C_P(\lambda, \beta) \tag{3.8}$$

where ρ is the density of the air, A is the swept area of the turbine blades, C_P is the power coefficient, β is the pitch angle of the blade, and λ is the tip speed ratio,

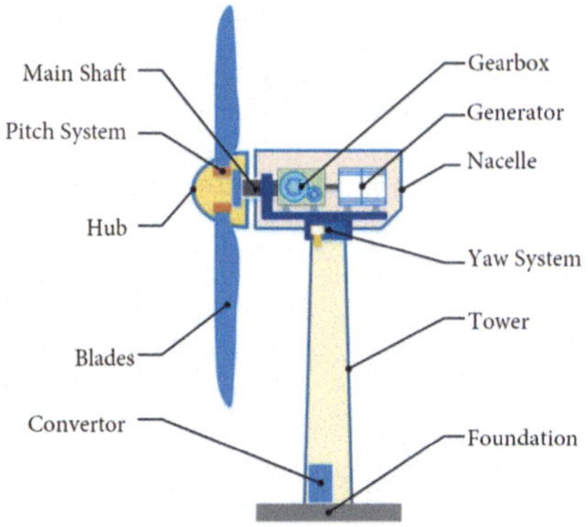

Fig. 3.9 Structure of the horizontal axis wind turbine

which relates the turbine speed ω_t to the wind speed v_v, given by:

$$\lambda = \frac{R_t \cdot \omega_t}{v_v} \tag{3.9}$$

where R_t is the radius of the turbine. The value of C_P depends on β and λ. Power coefficient curves are obtained experimentally or through simulations by wind turbine manufacturers. These curves affect the value of the extracted power, with C_P ranging between 0 and 0.59, a limit known as the Betz limit. The power coefficient C_P is given by:

$$C_P(\lambda, \beta) = 0.5176 \cdot (\frac{116}{\lambda_i} - 0.4 \cdot \beta - 5) \cdot e^{\frac{-21}{\lambda_i}} \tag{3.10}$$

where λ_i is given by:

$$\lambda_i = \frac{1}{\frac{1}{(\lambda+0.089)} - \frac{0.035}{\beta^3+1}} \tag{3.11}$$

Figure 3.9 illustrates the structure of a horizontal axis wind turbine (HAWT), which includes components such as blades, gearbox, generator, nacelle,[2] tower, yaw control systems, and blade pitch angle control β [33].

Figure 3.10 illustrates the curves of C_P as a function of λ for different values of β. It is observed that increasing the value of β reduces the value of C_P, with its maximum occurring at $\beta = 0°$ [34]. The power of the wind turbine is given by (3.8), and knowing the speed, the torque T_t applied to the generator shaft is obtained by:

$$T_t = \frac{P_t}{\omega_t} \tag{3.12}$$

[2] A protective structure at the top of the tower that houses the turbine generator.

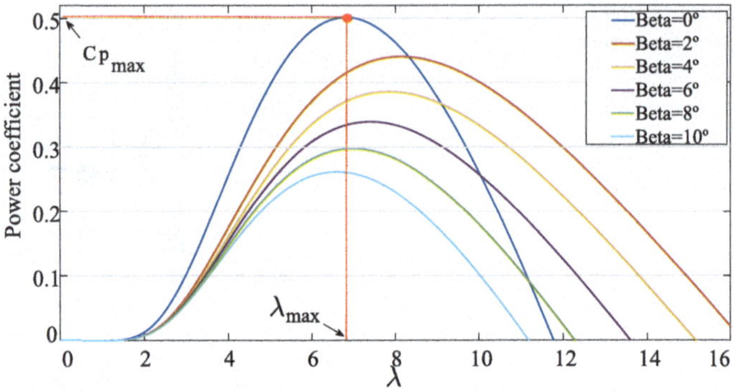

Fig. 3.10 Power coefficient × tip speed ratio

Fig. 3.11 Two-mass coupled model functioning as a gearbox

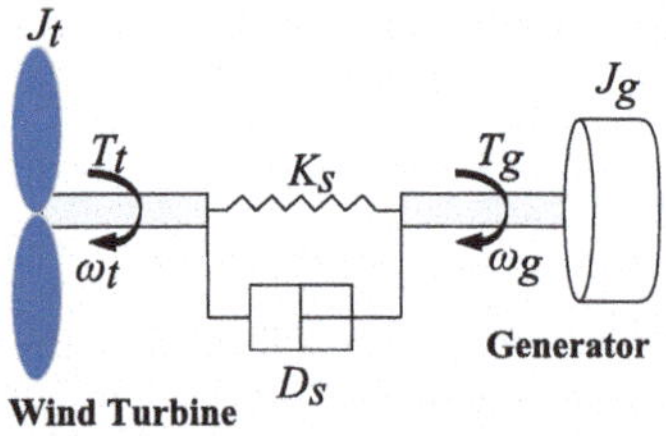

Various electric generator technologies are used in wind power generation, each with its own advantages and disadvantages. Generally, these technologies operate at higher speeds than conventional wind turbines. For example, induction generators operate in a higher speed range than wind turbines, while permanent magnet generators operate at lower speeds but with higher torque. Additionally, some generator technologies, such as the switched reluctance generator, offer a wide operating speed range.

Since the rotational speed of a wind turbine rotor is generally low compared to the synchronous speed of the electric generator, gearboxes are used to increase the rotational transmission ratio between the turbine rotor and the generator. These gearboxes act as speed multipliers, with various models traditionally employed. For example, there is the two-mass model, in which one mass represents the wind turbine and the other represents the generator. The two masses are connected by a shaft with modeled stiffness and damping, as illustrated in Fig. 3.11 [29].

In Fig. 3.11, T_t is the wind turbine torque, T_g is the generator torque, J_t is the inertia of the turbine, J_g is the inertia of the generator, ω_t is the turbine speed, ω_g is the generator speed, K_s is the stiffness of the shaft, and D_s is the damping constant. The gearbox increases the rotational transmission ratio, converting the rotor's rotational speed into a higher rotational speed suitable for generator operation.

References

1. S. Tarantola et al., Estimating the approximation error when fixing unessential factors in global sensitivity analysis. Reliab. Eng. Syst. Safety **92**(7), 957–960 (2007)
2. G.J. Klir, *Facets of Systems Science*, vol. 7 (Springer Science & Business Media, Heidelberg, Alemanha, 2013)
3. G.E. Mobus, M.C. Kalton, *Principles of Systems Science* (Springer, New York, 2015)
4. A. Hieronymi, Understanding systems science: a visual and integrative approach. Syst. Res. Behav. Sci. **30**(5), 580–595 (2013)
5. L. Von Bertalanffy, *General System Theory: Foundations, Development, Applications* (Revised Edition) (Georgetown University Library, Georgetown, 1969)
6. M.P. Wilson, General system theory: towards the unification of science, in *Proceedings of the 58th Annual Meeting of the ISSS-2014 United States*, vol. 1 (2015)
7. M. Newman, *Networks: An Introduction* (Oxford University Press, Oxford, 2010)
8. M. Bunge, *Emergence and Convergence: Qualitative Novelty and the Unity of Knowledge* (University of Toronto Press, Toronto, 2003)
9. H.A. Simon, The architecture of complexity. Proc. Am. Philos. Soc. **106**(6), 467–482 (1962)
10. V.M. Gomes et al., Mechanism for measuring system complexity apply ing sensitivity analysis. Complexity **2019**(1), 1303241 (2019)
11. G.A. Wainer, *Discrete-Event Modeling and Simulation: A Practitioner's Approach* (CRC Press, Boca Raton, FL, 2009)
12. R.D. Stacey, *Strategic Management and Organisational Dynamics: The Challenge of Complexity to Ways of Thinking About Irganisations* (Pearson Education, London, 2007)
13. J.R.B. Paiva et al., Complexity measure based on sensitivity analysis applied to an intensive care unit system. Sci. Reports **13**(1), 14602 (2023)
14. J.H. Holland, *Complexity: A Very Short Introduction* (OUP Oxford, Oxford, 2014)
15. M. Sayed-Mouchaweh, *Discrete Event Systems: Diagnosis and Diagnosability* (Springer Science & Business Media, Heidelberg, Alemanha, 2014)
16. C. Graham, D. Talay, *Stochastic Simulation and Monte Carlo Methods: Mathematical Foundations of Stochastic Simulation*, vol. 68 (Springer Science & Business Media, Heidelberg, Alemanha, 2013)
17. R.K. Arora, *Optimization - Algorithms and Applications* (CRC Press, Boca Raton, FL, 2015)
18. W.P. Calixto et al., Calculation of soil electrical conductivity using a genetic algorithm. Comput. Electron. Agri. **71**(1), 1–6 (2010)
19. M.W. Maier, E. Rechtin, *The Art of Systems Architecting*, vol. 1 (CRC Press, Boca Raton, FL, 2000)
20. V.M. Gomes et al., Analytical method for calculating the sensitivity index of system parameters, in *IEEE Internation Conference* (2017)
21. F.A. Gomes et al., Heuristic regression method for descriptive data analysis, in *2016 IEEE 16th International Conference on Environmentand Electrical Engineering (EEEIC)* (IEEE, 2016), pp. 1–5
22. W.P. Calixto et al., Parameters estimation of a horizontal multilayer soil using genetic algorithm. IEEE Trans. Power Delivery **25**(3), 1250–1257 (2010)
23. W.P. Calixto et al., Troubleshooting in geoelectrical prospecting using real-coded genetic algorithm with chromosomal extrapolation. Int. J. Numer. Modell. Electron. Networks Dev. Fields **28**(1), 78–95 (2015)
24. Z. Michalewicz, Heuristic methods for evolutionary computation techniques. J. Heuristics **1**, 177–206 (1996)
25. M.R.C. Reis et al., Nonlinear simulation methodology for switched reluctance machine using induction profile found by parametric regression, in *2017 CHILEAN Conference on Electrical, Electronics Engineering, Information and Communication Technologies (CHILECON)* (2017)
26. S. Koziel, X.-S. Yang, *Computational Optimization, Methods and Algorithms*, vol. 356 (Springer, New York, 2011)

27. M.R. Cunha Reis et al., Optimized techniques for driving and control of the switched reluctance motor to improve efficiency. Control Eng. Pract. **90**, 1–18 (2019)
28. W.P. Calixto et al., Troubleshooting in geoelectrical prospecting using real-coded genetic algorithm with chromosomal extrapolation. Int. J. Numer. Modell. Electron. Networks Dev. Fields **28**(1), 78–95 (2015). ISSN: 1099–1204. https://doi.org/10.1002/jnm.1986
29. L.D.S. Morais et al., Enhancement of the performance of switched reluctance generators in low wind speed conditions using advanced tracking techniques. Int. J. Energy Res. **2024**(1), 5541150 (2024)
30. D.E. Golberg, *Genetic algorithms in search, optimization, and machine learning*. Addion Wesley **1989**(102), 36 (1989)
31. M.O.L. Hansen, *Aerodynamics of Wind Turbines: Second Edition* (Earthscan, London, UK, 2008). ISBN: 978-1-84407-438-9
32. E. Hau, *Wind Turbines: Fundamentals, Technologies, Application, Economics* (Springer Science & Business Media, Heidelberg, Alemanha, 2013)
33. S. Wang et al., Dynamic analysis of wind turbines including nacelle–tower–foundation interaction for condition of incomplete structural parameters. Adv. Mech. Eng. **9**(3), 168781401769294 (2017)
34. A.B. Asghar, X. Liu, Estimation of wind turbine power coefficient by adaptive neuro-fuzzy methodology. Neurocomputing **238**, 227–233 (2017)

Chapter 4
Parametric Regression and Sensitivity Analysis

4.1 Parametric Regression

There is a need to represent physical systems using mathematical functions [1]. Regression models are used in a number of fields to express the relationship between variables within a given system. Some applications include: (i) business cycle estimation, (ii) medical treatment estimates, (iii) image processing, (iv) analysis of concrete structures, (v) audio processing, (vi) electric power systems, (vii) electrical drives, (viii) process control systems, and (ix) others [2, 3]. The regression process involves studying the relationship between input and output variables to describe them and develop a model for estimates and predictions of the physical system [1, 4]. Pilot plant tests are often conducted to identify the set of inputs and outputs that characterize the system, especially when the model or its parameters that describe the behavior of the system are unknown [2]. This process can involve significant effort, which in some cases renders system analysis using conventional methods impractical [5, 6].

Real or simulated systems provide discrete data as output; however, there is often a need to obtain a continuous mathematical function that describes the data [1]. Once the mathematical expression, or its parameters, representing the system is found, a number of analyses can be performed, including prediction. Prediction involves defining model outputs using inputs that exceed the system's maximum and minimum conditions. In such cases, the physical system is restricted by certain limits, and the regression model predicts outcomes beyond these limits while ensuring the integrity of the physical system [2, 7]. To address simulation challenges, computational numerical models can be used. However, these models often require significant computational time, especially when high levels of nonlinearity are present, complicating the analysis process and the implementation of new techniques. Regression can be used to find an analytical expression that represents the real system or part of it, integrating it into the model to reduce computational effort and improve the accuracy of the result [7, 8].

W. P. Calixto et al., *Control and Tracking Techniques for Switched Reluctance Machines*, Power Systems, https://doi.org/10.1007/978-3-031-86727-9_4

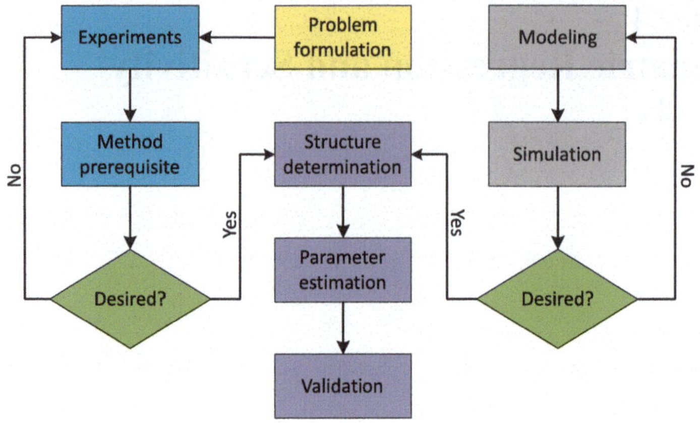

Fig. 4.1 Flowchart of the system identification procedure

Fig. 4.2 Diagram of the parametric regression optimization process

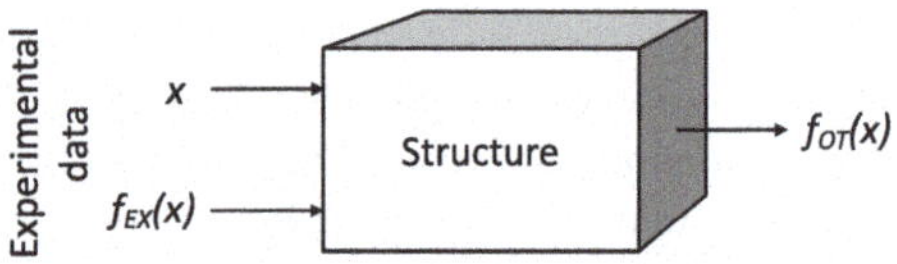

System identification techniques are typically integrated with optimization processes that involve various measures to fit candidate mathematical models to the real process. The selection of models and parameter adjustment are influenced by several factors, including: (i) prior knowledge of the system (linearity, degree of nonlinearity, delays), (ii) model properties (complexity), (iii) choice of the error metric to be minimized, and (iv) the presence of noise [9]. Figure 4.1 illustrates the diagram for the basic procedure for system identification.

In Fig. 4.1, it can be seen that regardless of previous knowledge of the system, parameter estimation is an important part of the identification process for subsequent validation. At this stage, optimization techniques are commonly used to obtain expressions that represent the real model. In addition to polynomial approximations that are typically used to represent the behavior of a given function or system, other types of approximations, such as exponential and trigonometric series, can be used [2]. The series chosen to represent a given system from experimental data should be selected based on the system's behavior and the need for extrapolation beyond the input value limits used in the test. Figure 4.2 illustrates the process of obtaining an expression for a given system from experimental data.

In Fig. 4.2, the experimental values x correspond to the input variables of the system, and $f_{EX}(x)$ represents the curve defined by the points experimentally obtained. The expression $f_{OT}(x)$ is the optimized representation of $f_{EX}(x)$, where the structure parameters are generated using optimization techniques to achieve values that bring the experimental and optimized curves closer together, i.e. $f_{EX}(x) \approx f_{OT}(x)$ [2]. This approximation is based on the evaluation function used in the optimization process, which typically involves minimizing the difference between the curves $f_{EX}(x)$ and $f_{OT}(x)$. In the structure shown in Fig. 4.2, the function $f_{OT}(x)$ can be composed of mathematical elements that adequately describe the system, such as polynomials, exponential expressions, and trigonometric series, including:

$$f_{OT} = a_0 + \sum_{i=1}^{n} a_i \cdot x^{b_i} \cdot \cos(c_i \cdot x + d_i) \cdot e^{e_i \cdot x} \tag{4.1}$$

where a_0, a_i, b_i, c_i, d_i, and $e_i \in \mathbb{R}$. Regression methods employ techniques that aim to achieve flexibility and predictive capacity. Numerous studies in the literature rely on polynomials or trigonometric functions for data approximation. Hybrid regressions, combining polynomial and trigonometric functions, offer greater representativeness, overcoming the limitations of non-periodicity in polynomials or excessive periodicity in trigonometric series [1, 2]. The regression process for identifying and fitting a model to represent the real system can be carried out either empirically or through simulation. The empirical approach, known in the literature as model synchronization, involves intuitively comparing the model outputs with the real system outputs. In contrast, the optimized regression process uses the output data from both the model and the real system as a metric for optimization [3, 7]. The regression process is illustrated in Fig. 4.3, where the region enclosed by the dashed rectangle represents the synchronization of the model [4].

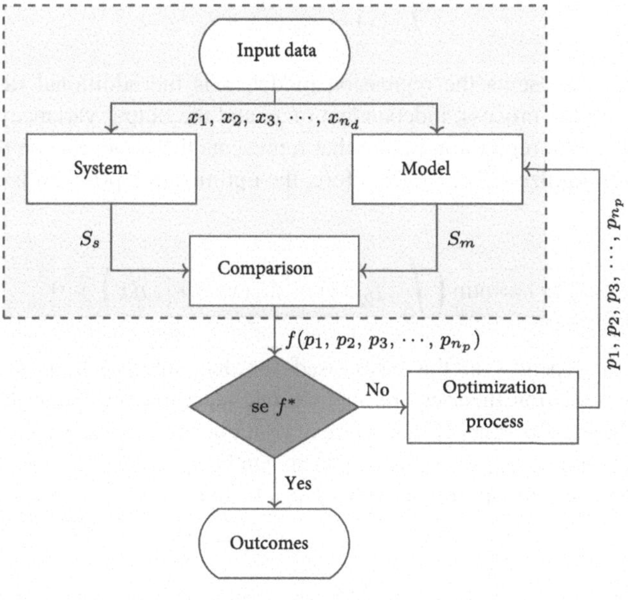

Fig. 4.3 Flowchart of the parametric regression process

4.1.1 Optimization Process in Parametric Regression

In Fig. 4.3, x represents the n_d inputs of both the real system and the model, where S_s and S_m are the outputs, respectively, and p are the adjustment parameters of the model, with n_p parameters [4]. The optimization algorithm adjusts the parameters p until the stopping criterion, based on the evaluation function, is met. Parametric regression is used as a mathematical modeling technique for model adjustment when there is little or no prior knowledge about the system model [10]. In the identification process, the mathematical model is obtained from data acquisition of the real system, allowing it to reproduce both static and dynamic characteristics. To use regression in system studies, mathematical modeling of the system is required for simulator implementation [1, 2]. Even in systems with bounded input and bounded output (BIBO) behavior, parametric regression can be used for prediction. A BIBO system is considered stable if a bounded control input always produces a bounded output [11]. Using the space H_p, we have:

$$1 \leq p < \infty : \; H_p = \left(f(t) \,\middle|\, \sqrt{\int_0^{\infty} |f(t)|^p \cdot dt} \right) < \infty \tag{4.2}$$

where $1 \leq p < \infty$ represents the subinterval of real numbers, and $f(t)$ is an integrable function within $1 \leq p < \infty$. Analyzing the experimental data $f_{ex}(x)$ from the BIBO system, the data obtained can be represented by:

$$f_{ex}(x) = f_{op}(x) + \epsilon \tag{4.3}$$

where $f_{op}(x)$ represents the regression model, ϵ is the additional random error associated with the process, independent of x, and the output variance is constant. Thus, $f_{op}(x)$ is the regression model that represents the system when the absolute error area is minimized [7, 11]. Therefore, the optimization problem is represented by:

$$f(x) = \min_{x \in \Omega} \left(\int_{\Omega} \left| f_{op}(x) - f_{ex}(x) + \epsilon \right| \cdot dx \right) > 0 \tag{4.4}$$

where $f_{op}(x)$ depends on the basis used for data interpolation. To solve this problem, interpolation theories are applied, one such method being the Legendre polynomial approximation [2]. The space of continuous functions $C_{[a,b]} \subset H_p[a, b]$ for any function $f \in C_{[a,b]}$, where $a, b \in \Re$, can be approximated by a polynomial function. Thus, any function defined by f can be expressed as a power series, such as:

$$\beta_1(x) = \left(1, p \cdot x, \ldots, p_n \cdot x^n\right) \tag{4.5}$$

where p represents the adjustment parameters of the function, and n is the number of bases to be used. Standard methods range from polynomial to trigonometric representations, using the base β_1 for the power or polynomial series and the base β_2 for the trigonometric series given by:

$$\beta_2(x) = \left[1, \sin\left(\frac{\pi \cdot x}{p}\right), \cos\left(\frac{\pi \cdot x}{p}\right), \ldots, \sin\left(\frac{n \cdot \pi \cdot x}{p_n}\right), \cos\left(\frac{n \cdot \pi \cdot x}{p_n}\right)\right] \tag{4.6}$$

The approximation of the results represents the system through an analytical expression, and regression methods are chosen on the basis of the problem's characteristics. The bases β_1 and β_2 have properties for representing continuous function spaces on the interval $[a, b]$. In cases where oscillations occur in the results, the base β_1 may be insufficient for polynomial regression over the interval. To capture these oscillations, trigonometric bases must be incorporated into the regression model [2]. The following model types can be applied: (i) white box, (ii) black box, and (iii) gray box. In a white-box model, knowledge of the system is required, as the modeling is conceptually performed. The black-box model has little to no prior knowledge of the system, and modeling is done empirically or through system identification techniques. The gray-box model involves modeling using information that is not part of the analyzed dataset within the system [1].

4.2 Parameterization of a Direct Current Motor as an Example

The methodology for identifying unknown parameters, such as in the case of a direct current (DC) motor, can be developed to determine electromechanical parameters. Experimental data are collected on the behavior of the motor's speed ω, armature current I_a, and induced electromotive force E_g over time t [12, 13]. In this scenario, the mathematical model of the DC motor is known and represented by:

$$\left.\frac{\omega(s)}{V_a(s)}\right|_{T_l=0} = \frac{K_t}{L_a \cdot J \cdot s^2 + (L_a \cdot B + R_a \cdot J) \cdot s + R_a \cdot B + K_t \cdot K_v} \tag{4.7}$$

where:

$$I_a(s) = \frac{V_a - L_a \cdot I_a \cdot s - E_g}{R_a} \tag{4.8}$$

and,

$$E_g(s) = K_v \cdot \omega \tag{4.9}$$

where R_a, L_a, K_t, K_v, B, and J are the electromechanical parameters defining the behavior of the DC motor, typically obtained by laboratory tests. The input V_a is applied to the system and the model, and the outputs ω, I_a, and E_g are observed so that the regression algorithm can compare these outputs and adjust the electromechanical parameters in the DC motor model [14].

4.3 Sensitivity Analysis

In system studies, varying scenarios and circumstances create the need to simplify models, identify errors, and most importantly, establish the relationship between the input and output variables of the system [15, 16]. In addition to system identification techniques, which allow the expressions to be obtained that represent the dynamic behavior of a given system, it is important to quantitatively understand the influence of input variables on the output variables of the system saltelli2004. Sensitivity analysis can be used, using various techniques, to quantify this influence. Sensitivity analysis studies the effect of a given input on a specific output, evaluating how output uncertainties can be attributed to different sources of input uncertainty. Sensitivity analysis can also be understood as the effort to quantify the relative contribution of each input to the model output [17, 18]. Sensitivity analysis is often accompanied by uncertainty analysis. The necessary difference between them is that the uncertainty analysis assesses the uncertainty of a given conclusion, while the sensitivity analysis identifies the sources of that uncertainty [19, 20].

To simulate a given system, it is necessary to find a model that adequately represents it while preserving its required characteristics [21]. The size of the model should not be confused with its complexity. In this context, sensitivity analysis can help identify the complexity of the model by verifying whether it processes parameters correctly and limiting complexity by fixing less relevant input variables, thus reducing computational effort during model simulation [22, 23]. Beyond identifying errors and simplifying the model, sensitivity analysis can be useful in detecting critical regions in the input space, evaluating policies, setting research priorities, and protecting against analyst bias or misinterpretation [24].

The sensitivity of the system to variations in input parameters can indicate: (i) research efforts needed to deepen the understanding of specific parameters, thus reducing output uncertainty, (ii) parameters that can be excluded from the final model due to their insignificance, (iii) input variables that contribute the most to output variability, (iv) parameters with the highest correlation to the output, and (v) the consequences of changes in specific input parameters once the model is in use [25]. Once the purpose of the analysis is defined, the most appropriate method for the sensitivity analysis is selected, as different methods may produce divergent responses to the same questions.

4.4 Methods of Sensitivity Analysis

Sensitivity analysis methods provide insight into the relationship between input and output variables in systems. Depending on the purpose of the analysis, the input parameters and outputs of interest are defined, which can be quantitative or qualitative. Input parameters refer to independent variables that can be manipulated, while outputs refer to dependent variables, which can be affected by changes in input parameters [26]. The manipulation of input parameters is carried out using sensitivity analysis methods, classified according to the approach adopted regarding the input parameter search space: local or global. Local analysis depends on point estimates of parameter values, while global analysis examines sensitivity based on the entire distribution of the parameters [27]. The local approach evaluates the output of the system by varying the input parameters one at a time [28].

Thus, while one parameter varies, the others are kept fixed at their central (nominal) values. In global analysis, in addition to exploring the input parameter space within a finite (or feasible) region, outputs are evaluated globally by averaging the variation across all input parameters. Figure 4.4 illustrates measures related to a one-at-a-time (OAT) approach: while one parameter (indicated by the red arrow) has altered its values, the others remain fixed at the nominal or baseline value, $\alpha = (\alpha_1, \alpha_2, \cdots, \alpha_n)$. In the first cycle, only the value of the parameter x_1 is changed, in the second cycle, the value of x_2 is modified, and so on, until the n-th cycle, where only the value x_n is varied.

Although limited to the central point (or baseline case), local sensitivity analysis methods are frequently used due to their relatively low computational effort, unlike global analysis, which typically requires simulations of the order of 10^4 [29, 30]. In the sensitivity analysis process, global methods require time-consuming steps, such as: (i) defining the uncertainty distributions of input parameters and (ii) simulating the system for various input sets. The time needed to determine the appropriate uncertainty distribution for each parameter can be financially burdensome, as it

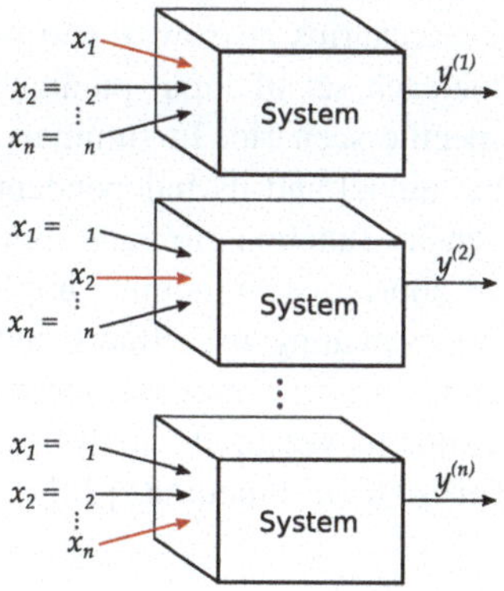

Fig. 4.4 Diagram of one-at-a-time measures for n input parameters

may involve accessing multiple sources to gather the necessary knowledge about input parameters, including expert opinions [24]. The time consumed during the simulation phase depends on the model and the hardware used for data processing in the simulator.

Assuming that the probabilistic distribution governing each input parameter is known, random values for the parameters of interest can be generated. Thus, for the input parameters n and the scenarios m (sets of input values), the matrix X is defined as shown in (4.10). In most cases, it is assumed that input parameters are independent of each other, allowing random values to be generated separately from each parameter's distribution. To obtain the output value corresponding to each set of input parameter values (rows of matrix X), the model simulation is performed m times, resulting in matrix Y, as shown in (4.11).

$$X = \begin{bmatrix} x_1^{(1)} & x_2^{(1)} & \cdots & x_n^{(1)} \\ x_1^{(2)} & x_2^{(2)} & \cdots & x_n^{(2)} \\ x_1^{(3)} & x_2^{(3)} & \cdots & x_n^{(3)} \\ \vdots & \vdots & \ddots & \vdots \\ x_1^{(m-1)} & x_2^{(m-1)} & \cdots & x_n^{(m-1)} \\ x_1^{(m)} & x_2^{(m)} & \cdots & x_n^{(m)} \end{bmatrix} \tag{4.10}$$

$$Y = \begin{bmatrix} y^{(1)} \\ y^{(2)} \\ y^{(3)} \\ \vdots \\ y^{(m-1)} \\ y^{(m)} \end{bmatrix} \tag{4.11}$$

where matrix X contains the values of the input parameters x_1 to x_n (columns) for scenarios m (rows), and matrix Y contains the corresponding output values y for each set of input parameters, called scenarios. The superscripts indicate the specific scenario. In summary, sensitivity analysis typically involves: (i) defining the model and its independent and dependent variables, (ii) assigning probability density functions to each input parameter, (iii) generating the input matrix using an appropriate random sampling method, (iv) calculating the output vector, and (v) evaluating the effects and the relative importance of the input–output relationship [25]. Once the input parameters are defined and the model or system is available, sensitivity analysis can be performed using: (i) visual, (ii) analytical, and (iii) statistical methods [22].

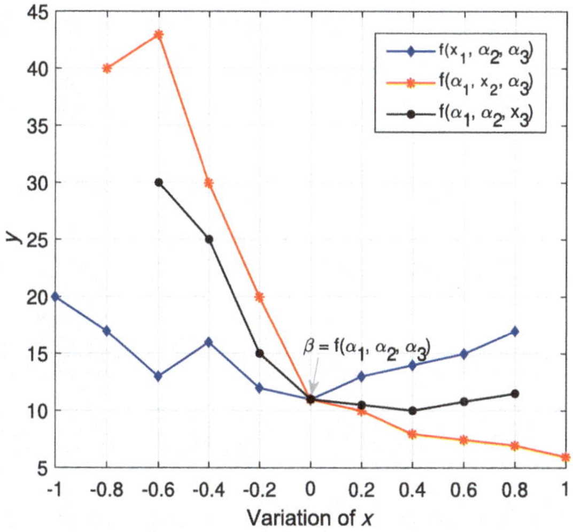

Fig. 4.5 Example of a spider plot with three parameters: x_1, x_2, and x_3

4.4.1 Visual Methods

Two visual methods are commonly used to analyze system sensitivity: spider plots and scatter plots. The first method involves output measurements for one-at-a-time parameter variations, while the second method uses the scatter of each parameter relative to the output values for different input sets. The spider plot shows the variation curves of the input parameters, as illustrated in Fig. 4.5 [31]. This variation can be negative or positive from the baseline value of the parameter α_j. When $x_1 = \alpha_1, x_2 = \alpha_2, \cdots, x_n = \alpha_n$, the system output β corresponds to $y = f(\alpha_1, \alpha_2, \cdots, \alpha_n)$. Graphically, the output β is represented by the central intersection point of the sensitivity curves.

The baseline case $\alpha = (\alpha_1, \alpha_2, \cdots, \alpha_n)$ usually refers to the optimal or optimized solution obtained after the optimization process or represents the best estimate of the inputs as defined by a system expert [15, 21]. Starting from the baseline value, the parameter can be varied up to $\pm$ 100% or within the limits of its feasible range. In practical situations, establishing the range of parameter variation involves defining operational ranges for the system without compromising its physical limits. Due to operational or optimization constraints, parameters may have restrictions on the extent of their variation range, showing only values close to the baseline [32]. Another graphical method for analyzing parameter sensitivity is through the scatter plot, often used after simulation of the model with probabilistic methods. The influence of the input parameters is individually assessed for a specific output, and the analysis is visual [22]. Taking into account the matrix X, defined by (4.10), which contains input sets m, the corresponding output set Y, defined by (4.11), with dimensions $m \times 1$, can be obtained. With these data, n scatter plots can be generated by plotting the m output values y against the m values of each of the n input parameters [24].

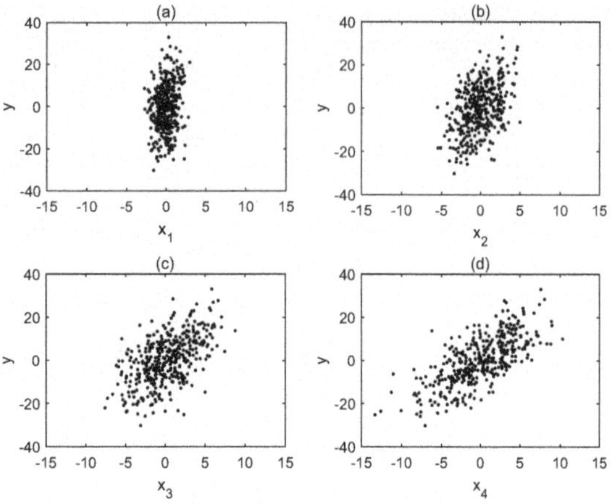

Fig. 4.6 Scatter plot of input parameters from x_1 to x_4 relative to the output y

For illustration, Fig. 4.6 shows scatter plots for a model with four parameters, $x_1, \ldots, x_4$, whose probabilistic distributions $N(\overline{\mu}x_j, \sigma)$ are normal distributions with means $\overline{\mu}x_j$ and standard deviation σ, given as: $N(0, 1)$, $N(0, 2)$, $N(0, 3)$, and $N(0, 4)$, respectively. Using Monte Carlo simulation, scenarios can be generated based on these distributions and the model can be simulated to obtain the set of outputs Y. Figure 4.6 illustrates the scatter of each parameter relative to the output y, where $y = f(x_1, \ldots, x_4)$.

It can be observed in Fig. 4.6 that the scatter indicates the variability of the parameter. Figure 4.6a corresponds to the parameter with the lowest variability, as all points are concentrated within the interval $[-5, 5]$, while Fig. 4.6d represents the parameter with the highest variability, where the scatter spans the interval $[-15, 15]$. In addition to variability, the shape or pattern of the points in the scatter plot should be noted. Figure 4.6d shows a scatter pattern with points indicating a trend, due to the correlation between input x_4 and output y. The system in question is more sensitive to variations in x_4, followed by x_3, x_2, and x_1, in the order [24].

Models exhibit sensitivity to input parameters in two distinct ways: (i) parameter variability or uncertainty, where input uncertainty propagates through the model, contributing to output variability, and (ii) correlation between input and output, where the output may be highly correlated with a specific input parameter [25]. Therefore, important parameters with high variability, as indicated by uncertainty analysis, are also sensitive parameters. However, there may be sensitive parameters with low variability that are considered less important by uncertainty analysis but can still lead to significant changes in the output. Thus, scatter plots can visually represent both variability and trends in the input–output relationship.

To visually express the local trend of the parameters, the generalized sensitivity scatter plot is used. In this type of plot, the sensitivity coefficients are visually represented as flow lines, showing the continuity and orientation of the data while addressing the issue of point overlap [33]. These coefficients are calculated using a method that involves three transformations: (i) subspace selection, (ii) differentiation with respect to the variable of interest, and (iii) projection. In various applications, visual methods provide an indication of how a given output is affected by input variations. These methods are typically used in preliminary sensitivity analysis studies or as a complement to the results of analytical and statistical methods [22].

4.4.2 *Analytical Methods*

Analytical methods provide mechanisms to quantify the sensitivity of the input–output relationship. In this context, derivative-based methods or factorial design are frequently applied in fields such as chemistry, physics, medicine, and engineering. In derivative-based methods, the one-at-a-time approach is used to calculate derivatives, whereas in factorial design, all parameters are varied simultaneously. The derivatives can mathematically define how sensitive the output y is with respect to the input parameter x_j. In one-at-a-time analysis, partial derivatives are commonly used and can be normalized by the nominal value of the parameter or its standard deviation [28]. Sensitivity is given by the partial derivative (denoted by the superscript p) of y with respect to x_j. From (4.12), (4.13) can be derived, where the sensitivity of parameter x_j results from the partial derivative normalized by the standard deviation of the input σ_{x_j} and the output σ_y [24].

$$S^p_{x_j} = \frac{\partial y}{\partial x_j} \tag{4.12}$$

$$S^\sigma_{x_j} = \frac{\sigma_{x_j} \partial y}{\sigma_y \partial x_j} \tag{4.13}$$

Derivative-based sensitivity analysis is restricted to the central point in the input parameter space and is unjustified when the input is uncertain and the model's linearity is unknown. To explore the input space, the local approach can be applied globally, for example, by calculating incremental rates such as $\frac{y(x_j+\Delta x_j)-y(x_j)}{\Delta x_j}$ at different points in the parameter space to obtain an average output for y [24]. However, only one parameter at a time is varied for each selected point. When the goal is to vary all parameters simultaneously and measure the resulting effect on the output due to these variations, the analytical method of factorial design can be applied. In its simplest form, the factorial design 2^n allows the input parameters n to be taken on two levels, resulting in 2^n possible combinations characterizing the input set. Table 4.1 presents all the combinations ν of levels (represented by $-$ and $+$) for parameters x_1, x_2, and x_3, together with their respective outputs $y^{(\nu)}$, where $1 \leq \nu \leq 2^3$ [34].

The sensitivity indices for the input parameters x_1, x_2, and x_3 are expressed by (4.14), (4.15), and (4.16), respectively, quantifying the effect of the level

Table 4.1 Factorial design combinations for 2^3

ν	x_1	x_2	x_3	$y^{(\nu)}$
1	−	−	−	$y^{(1)}$
2	+	−	−	$y^{(2)}$
3	−	+	−	$y^{(3)}$
4	+	+	−	$y^{(4)}$
5	−	−	+	$y^{(5)}$
6	+	−	+	$y^{(6)}$
7	−	+	+	$y^{(7)}$
8	+	+	+	$y^{(8)}$

variation. The generalization of (4.14), (4.15), and (4.16) results in (4.17), which corresponds to the difference between the mean output when the parameter is at level + and when the parameter is at level − [26, 34].

$$S_{x_1}^f = \frac{(y^{(2)} - y^{(1)}) + (y^{(4)} - y^{(3)}) + (y^{(6)} - y^{(5)}) + (y^{(8)} - y^{(7)})}{4} \tag{4.14}$$

$$S_{x_2}^f = \frac{(y^{(3)} - y^{(1)}) + (y^{(4)} - y^{(2)}) + (y^{(7)} - y^{(5)}) + (y^{(8)} - y^{(6)})}{4} \tag{4.15}$$

$$S_{x_3}^f = \frac{(y^{(5)} - y^{(1)}) + (y^{(6)} - y^{(2)}) + (y^{(7)} - y^{(3)}) + (y^{(8)} - y^{(4)})}{4} \tag{4.16}$$

$$S_{x_j}^f = \bar{y}_{x_j^+} - \bar{y}_{x_j^-} \tag{4.17}$$

The value of sensitivity $S_{x_j}^f$ can be positive or negative, indicating an average increase or decrease in output as x_j transitions from the level x_j^+ to the level x_j^-. In practice, these levels can be represented by quantitative or qualitative values of the variables under analysis, and the experimental design should be directed by the proposed objective [26]. The combinations depend on the number of input parameters and the levels associated with each parameter. In a factorial design 2^n, it is assumed that all parameters have only two levels. However, if the number of levels varies, the total number of combinations will equal the product of the number of levels of one parameter multiplied by that of the subsequent parameter, continuing this process until reaching the number of levels of the final input parameter.

Each combination of input parameter levels corresponds to an experiment that is typically repeated several times to estimate the experimental error. Therefore, all possible combinations of input parameter levels represent the minimum number required to conduct a complete factorial design. It is advisable to select the most relevant input parameters for analysis, using fractional factorial designs, for example [26]. In addition to derivatives and factorial designs, other analytical methods can be used to measure the sensitivity of input parameters, such as the method of elementary effects, nominal range sensitivity analysis, equilibrium point analysis, difference in ratio, and automatic differentiation, among others. These methods are limited to evaluating the impact of variation ranges in input values on output. Consequently, investigations are conducted into the relationship between the output variance and the input variance with the help of statistical methods [22].

4.4.3 Statistical Methods

Based on statistical measures of central tendency and variability, sensitivity analysis methods can explore the input parameter space and determine the relationships that define the influence of each parameter on the model output. Random value sets for input parameters can be generated, as shown in (4.10), allowing the computation of the output $y = f(x_1, x_2, \cdots, x_n)$ for each set of input values. By applying various techniques such as conditional variance and total effects, global sensitivity measures can be derived for the input parameters. In contrast to local analysis, which examines one parameter at a time, the conditional variance method involves measuring the variation of $n - 1$ parameters using their probabilistic distributions while keeping one parameter constant. The term conditional refers to the condition of fixing one input parameter and observing the variability of the output, comparing it to the unconditional or total variance, represented by (4.18), which is obtained when all input parameters n have variable values.

$$\sigma^2(Y) = \overline{\mu}_{\sigma^2}\left(Y|_{x_{\sim j}}\right) + \sigma^2_{\overline{\mu}}\left(Y|_{x_{\sim j}}\right) \tag{4.18}$$

where $x_{\sim j}$ represents the variation of all input parameters except for x_j, $\overline{\mu}\sigma^2$ is the mean of the variance, $\sigma^2\overline{\mu}$ is the variance of the mean, Y is given by (4.11), and $Y|x\sim j$ indicates the output vector when x_j remains constant, which means that x_j does not vary. The mean is used since the parameter under analysis assumes distinct values within its range of possibilities; If x_j were fixed at a single value, the analysis would depend on that point. Thus, the sensitivity of the parameter analyzed is determined by the ratio of the variance of the means of the output (for x_j fixed at different points) to the total variance, as given by (4.19), where the superscript σ^2 in S_{x_j} denotes the use of conditional variance to compute the sensitivity index of the input parameters.

$$S_{x_j}^{\sigma^2} = \frac{\sigma^2_{\overline{\mu}}\left(Y|_{x_{\sim j}}\right)}{\sigma^2(Y)} \tag{4.19}$$

Taking into account the sensitivity calculation in (4.19), the most sensitive parameter is defined as the one that, when fixed, results in the greatest average reduction in output variance. The term **on average** is used because the average is calculated from the distribution of the parameter itself [24]. Since the analysis focuses on one parameter alone, it is referred to as first-order sensitivity analysis. If the input parameters are taken two at a time (or more), the analysis is classified as second-order (or higher-order) sensitivity analysis, expressed as:

$$\sigma^2_{\overline{\mu}}\left(Y|_{x_{\sim j},x_{\sim l}}\right) = \sigma^2_{\overline{\mu}}\left(Y|_{x_{\sim j}}\right) + \sigma^2_{\overline{\mu}}\left(Y|_{x_{\sim l}}\right) + \sigma^2_{\delta_{x_j,x_l}}(Y) \tag{4.20}$$

where $j, l \in \mathbb{N}^*$ with $l \neq j \leq n$. The term $\sigma^2_{\delta_{x_j,x_l}}(Y)$ represents the variance of Y due to the interaction between the input parameters x_j and x_l, which contributes to the output y in a way that cannot be captured by the separate effects of x_j and x_l. Taking into account the effects arising from the interactions of the parameters, the **total effects method** includes all terms of any order that involve the parameter under analysis [24, 35]. For example, in a generic model with three input parameters, we have:

$$\begin{aligned}
S^{\tau}_{x_1} &= S_{x_1} + S_{x_1,x_2} + S_{x_1,x_3} + S_{x_1,x_2,x_3} \\
S^{\tau}_{x_2} &= S_{x_2} + S_{x_1,x_2} + S_{x_2,x_3} + S_{x_1,x_2,x_3} \\
S^{\tau}_{x_3} &= S_{x_3} + S_{x_1,x_3} + S_{x_2,x_3} + S_{x_1,x_2,x_3}
\end{aligned} \tag{4.21}$$

where $S^{\tau}_{x_1}$, $S^{\tau}_{x_2}$, and $S^{\tau}_{x_3}$ are the sensitivity indices that represent the total effects of x_1, x_2, and x_3, respectively. In (4.21), there are seven distinct interaction terms: S_{x_1}, S_{x_2}, S_{x_3}, S_{x_1,x_2}, S_{x_1,x_3}, S_{x_2,x_3}, and S_{x_1,x_2,x_3}. The formation law is given by $2^n - 1$, where n represents the number of input parameters. Generally, (4.21) is equivalent to (4.22), defined as the difference between the sum of all possible sensitivity terms and the first-order sensitivity of the parameter, obtained through the conditional variance method. Thus, considering (4.18), this difference corresponds to the relationship between the mean of the conditional variances and the total variance [24].

$$S^{\tau}_{x_j} = 1 - \frac{\sigma^2_{\overline{\mu}}\left(Y|_{x_{\sim j}}\right)}{\sigma^2(Y)} = \frac{\overline{\mu}_{\sigma^2}\left(Y|_{x_{\sim j}}\right)}{\sigma^2(Y)} \tag{4.22}$$

where $S^{\tau}_{x_j}$ is the sensitivity index of x_j calculated based on total effects. Variance-based methods can lead to a comprehensive understanding of the sensitivity pattern of the model when all interaction terms are considered, specifically $2^n - 1$. However, it is sufficient to characterize the sensitivity pattern through the total set of first-order terms and total effects. In the case of analyzing the factors 10, the sensitivity terms 20 are calculated instead of 1023. In addition to the conditional variance and total effects methods, statistical methods also include regression analysis, response surface methods, Fourier amplitude sensitivity testing, mutual information index, among others [22].

4.4.4 Area Method

The area method in a system with n input parameters, represented by $x_1, x_2, \cdots, x_i, \cdots, x_n$, is illustrated in Fig. 4.5. This sensitivity analysis is local and investigates the reference values of the parameters that lead the system to its optimal performance [36]. The set of these reference values, called the base case $\overrightarrow{\beta}$, is defined as:

$$\overrightarrow{\beta} = [x_1^\beta, x_2^\beta, \cdots, x_i^\beta, \cdots, x_n^\beta] \quad (4.23)$$

where x_i^β denotes the reference value of parameter x_i. After defining the base values of the parameters, a set of values must be chosen for each parameter within the variation range of -100% to 100% relative to the base value. During the analysis, while one parameter (the parameter under analysis) is modified, the others remain constant at their reference values. This process enables the assessment of the system's sensitivity to variations in each parameter x_i. This analysis involves defining multiple scenarios where different values of the parameter under analysis within the variation range are combined with the reference values of the other parameters. These scenarios are simulated, and the resulting output is stored in $\overrightarrow{Y_i}$ [15].

To evaluate the results of the sensitivity analysis using the area method, all parameters are assumed to vary within the range of -100% to 100% relative to the base value $\overrightarrow{\beta}$. However, in practical situations, some parameters may have physical limitations that restrict their range of variation. The more sensitive a parameter is (considering the difference between the obtained output value and the output value corresponding to the scenario $\overrightarrow{\beta}$, the greater its impact on the output of the system. Based on this evaluation, the area method is used to calculate the sensitivity indices of the input parameters. The area method relies on the relationship between the areas defined by the parameter curves. The parallel axis to the abscissa in the spider diagram is called the base axis, as shown in Fig. 4.7 [15, 36].

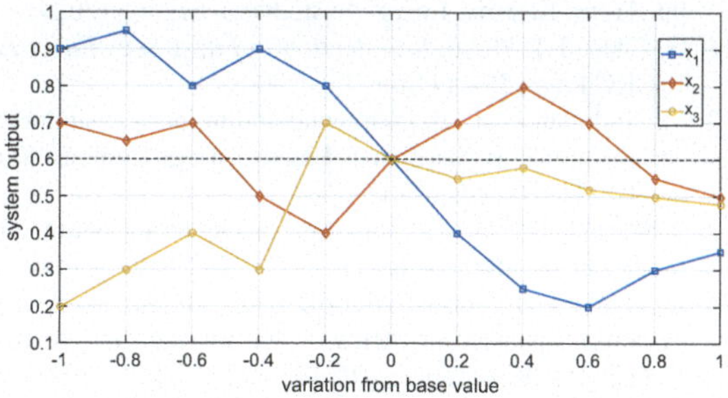

Fig. 4.7 Spider diagram of the system sensitivity analysis with three input parameters

The areas between each parameter variation curve and the base axis represent the parameter's contribution to the total area. The sensitivity index of a parameter is calculated using Eq. (4.24), where $S^a_{x_i}$ is the sensitivity index of the parameter x_i, A_{x_i} is the area defined by the variation curve of parameter x_i and the base axis, and n is the number of input parameters [15, 37].

$$S^a_{x_i} = \frac{A_{x_i}}{\sum_{i=1}^{n} A_{x_i}} \tag{4.24}$$

References

1. K.J. Keesman, *System Identification: An Introduction* (Springer Science & Business Media, Heidelberg, Alemanha, 2011)
2. F.A. Gomes et al., Heuristic regression method for descriptive data analysis, in *2016 IEEE 16th International Conference on Environment and Electrical Engineering (EEEIC)* (IEEE, 2016), pp. 1–5
3. M.R. Cunha Reis et al., Optimized techniques for driving and control of the switched reluctance motor to improve efficiency. Control Eng. Pract. **90**, 1–18 (2019)
4. A.H.F. Silva et al., Parametric regression applied for determination of electrical parameters of synchronous and induction generators operating in parallel on the electrical energy repowering system. Energies **14**(13), 3875 (2021)
5. R. Krishnan, *Eletric Motor Drives: Modeling, Analysis, and Control*, vol. 1 (Prentice Hall, New Jersey, 2001)
6. N.S. Nise, *Control Systems Engineering* (Wiley, New York, 2020)
7. M.R.C. Reis et al., Increasing efficiency of the switched reluctance generator using parametric regression and optimization methods, in *2017 18th International Scientific Conference on Electric Power Engineering (EPE)* (IEEE, 2017), pp. 1–6
8. D.A. Andrade, R. Krishnan, Characterization of switched reluctance machines using Fourier series approach, in *Conference Record of the 2001 IEEE Industry Applications Conference. 36th IAS Annual Meeting (Cat. No. 01CH37248)*, vol. 1 (IEEE, 2001), pp. 48–54
9. R. Johansson, *System Modeling and Identification* (Prentice Hall, New Jersey, 1993). ISBN: 0134823087
10. A.S. Magalhaes et al., Experimental study of induction generator as a re-powering solution. Int. Trans. Electric. Energy Syst. **30**(6), e12365 (2020)
11. G. Chen, T.T. Pham, *Introduction to Fuzzy Sets, Fuzzy Logic and Fuzzy Control Systems*, vol. 1 (CRC Press, Boca Raton, FL, 2000)
12. M. Abdullah et al., Experimental and numerical studies for parameters identification of direct current motor. Indonesian J. Electric. Eng. Comput. Sci. **27**(2), 592–600 (2022). https://doi.org/10.11591/ijeecs.v27.i2
13. A.M.-C. Felipe et al., Estimation of DC motor parameters using the least square estimator. J. Math. Quant. Methods **6**(11), 18–23 (2022). https://doi.org/10.35429/jmqm.2022.11.6.18.23
14. K. Nguyen, S. Vlasov, Algorithm for identification of DC motor parameters by method of dynamic expansion of regressor and mixing. Sci. Tech. J. Inf. Tech. Mech. Opt. **21**(1), 130–134 (2021). https://doi.org/10.17586/2226-1494-2021-21-1-130-134

15. J.P. Saraiva, B.S. Lima, V.M. Gomes, P.H.R. Flores, F.A. Gomes, A.O. Assis, Calculation of sensitivity index using one-at-a-time measures based on graphical analysis, in *2017 18th International Scientific Conference on Electric Power Engineering (EPE)* (May 2017), pp. 1–6. https://doi.org/10.1109/EPE.2017.7967329
16. R.L. Iman, J.C. Helton, An investigation of uncertanty and sensitivity analysis techniques for computer-models. Risk Anal. **8**(1), 71–90 (1988)
17. T. Homma, A. Saltelli, Importance measures in global sensitivity analysis of nonlinear models. Reliab. Eng. Syst. Safety **52**(1), 1–17 (1996)
18. S. Marino et al., A methodology for performing global uncertainty and sensitivity analysis in systems biology. J. Theoret. Biol. **254**(1), 178–196 (2008)
19. A. Saltelli, P. Annoni, How to avoid a perfunctory sensitivity analysis. Environ. Modell. Software **25**(12), 1508–1517 (2010)
20. D.P. Loucks, E.V. Beek, *Water Resources Systems Planning and Management* (UNESCO, Paris, 2005)
21. D.J. Pannel, *Sensitivity Analysis: Strategies, Methods, Concepts, Examples* (University of Western Australia, Crawley, 2009)
22. H.C. Frey, S.R. Patil, Identification and review of sensitivity analysis methods. Risk Anal. **22**(3), 553–578 (2002)
23. R.L. Iman, W.J. Conover, Small sample sensitivity analysis techniques for computer-models, with an application to risk assessment. Commun. Stat. A Theory Methods **9**(17), 1749–1842 (1980)
24. A. Saltelli et al., *Global Sensitivity Analysis: The Primer* (Wiley, New York, 2008)
25. D.M. Hamby, A review of techniques for parameter sensitivity analysis of environmental models. Environ. Monitor. Assess. **32**(2), 135–154 (1994)
26. O. Andersson, *Experiment!: Planning, Implementing and Interpreting* (Wiley, New York, 2012)
27. D.M. Hamby, A comparison of sensitivity analysis techniques. Health Phys. **68**(2), 195–204 (1995)
28. A. Saltelli, S. Tarantola, K.P.S. Chan, A quantitative model-independent method for global sensitivity analysis of model output. Technometrics **41**(1), 39–56 (1999)
29. H.M. Griffiths, M.C. Kennedy, P. Ioannou-Katidis, G.P. Petropoulos, G. Ireland, D.P. Kalivas, Extending the global sensitivity analysis of the Sim-Sphere model in the context of its future exploitation by the scientific community. Water **7**(5), 2101–2141 (2015)
30. J.P. Rivera, J. Muñoz-Marí, J. Vicent, G. Camps-Valls, J. Verrelst, N. Sabater, J. Moreno, Emulation of leaf, canopy and atmosphere radiative transfer models for fast global sensitivity analysis. Remote Sens. **8**(8), 673 (2016)
31. T.G. Eschenbach, L.S. McKeague, Exposition on using graphs for sensitivity analysis. Eng. Econom. **34**(4), 313–333 (1989)
32. V.M. Gomes et al., Analytical method for calculating the sensitivity index of system parameters, in *2017 CHILEAN Conference on Electrical, Electronics Engineering, Information and Communication Technologies (CHILECON)* (IEEE, 2017), pp. 1–6
33. Y.-H. Chan, C.D. Correa, K.-L. Ma, The generalized sensitivity scatterplot. IEEE Trans. Visual. Comput. Graph. **19**(10), 1768–1781 (2013)
34. A.M. Law, W. David Kelton, *Simulation Modeling and Analysis*, vol. 3 (McGraw-Hill, New York, 2007)
35. F. Sarrazin, F. Pianosi, T. Wagener, Global sensitivity analysis of environmental models: convergence and validation. Environ. Modell. Software **79**, 135–152 (2016)
36. V.M. Gomes et al., Mechanism for measuring system complexity applying sensitivity analysis. Complexity **2019**(1), 1303241 (2019)
37. J.R.B. Paiva et al., Stability metric based on sensitivity analysis applied to electrical repowering system. Energies **14**(22), 7824 (2021)

Chapter 5
Design and Development of the Experimental Platform

5.1 Construction of Experimental Platform

To facilitate a comprehensive analysis of the behavior of the switched reluctance machine (SRM) operating as a motor, it is necessary to apply mechanical torque to its shaft, allowing evaluation of its performance during activation and control [1]. The system illustrated in Fig. 5.1 shows the SRM with three spools $\varsigma' = 3$, featuring six poles in the stator and four in the rotor (6×4), coupled with a three-phase induction machine (IM), which, in this case, functions as an electromagnetic brake [2].

The notations L_1', L_2', L_3', and L_1'', L_2'', L_3'' represent the terminals of the windings ς of the switched reluctance machine (SRM), while U, V, W, U', V', and W' denote the terminals of the windings of the three-phase induction machine (IM). For SRM operation, it is necessary to know the position of the rotor θ to energize the coils during phases when the phase inductance increases or decreases, thus providing positive or negative torque (motor operation and generation operation, respectively). Information about θ can be obtained through direct or indirect measurements. For direct measurement, an absolute optical encoder is used [3].

5.2 Switched Reluctance Generator Operation

When operating as a switched resistance generator (SRG), the SRM requires external mechanical power supplied by a primary machine, which applies the external mechanical torque T_M to the shaft of the SRG. Figure 5.2 illustrates the SRG and the primary machine for generator activation [3].

The assembly illustrated in Fig. 5.2 comprises a three-phase induction motor (IM), which provides external torque to the shaft of the switched reluctance generator (SRG). This primary machine is powered by a frequency converter to

W. P. Calixto et al., *Control and Tracking Techniques for Switched Reluctance Machines*, Power Systems, https://doi.org/10.1007/978-3-031-86727-9_5

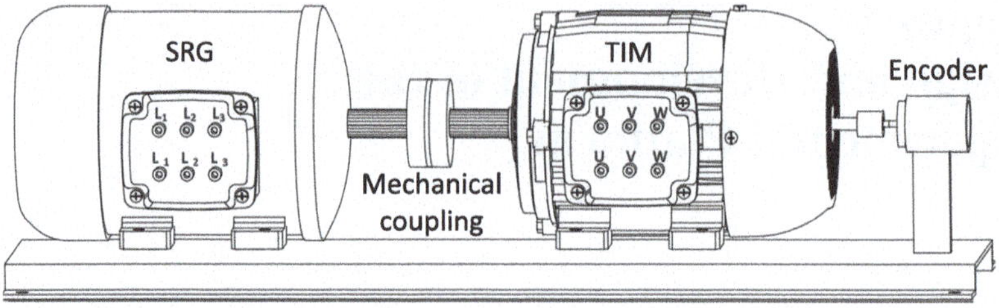

Fig. 5.1 Switched reluctance machine associated with a three-phase induction machine

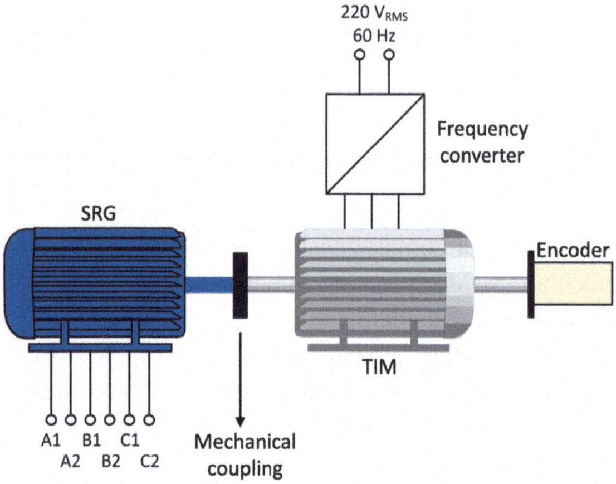

Fig. 5.2 Integration of the switched reluctance generator and primary machine for external torque supply

enable operation at various speeds, allowing the observation of the SRG's behavior in response to speed variations. Thus, the combination of the IM and the frequency converter serves the role of a wind turbine, providing mechanical power to the SRG. Mechanical coupling between machines requires the use of a position sensor. This is achieved by integrating an absolute encoder, as the operation of the SRG depends on the instantaneous angular position of the rotor θ, as expressed in (2.1) and shown in Fig. 2.4. The coupling of the SRG, IM, and encoder is illustrated in Fig. 5.2.

In Fig. 5.2, the terminals of the switched reluctance generator (SRG), labeled A1, A2, B1, B2, C1, and C2 correspond to the terminals of the three windings of the generator, reflecting its topology 6×4. These windings must be energized at the appropriate moments to enable their magnetization, followed by energizing the load connected to the generator terminals during the demagnetization phase. This process is facilitated by the use of a power converter, illustrated in Figs. 2.5, 2.6, and 2.8. Figure 5.3 shows the integration of the power converter with the SRG, the induction motor (IM), and the encoder shown in Fig. 5.2, where V_1 and V_2 represent the excitation voltage and the output voltage across the load, respectively [2].

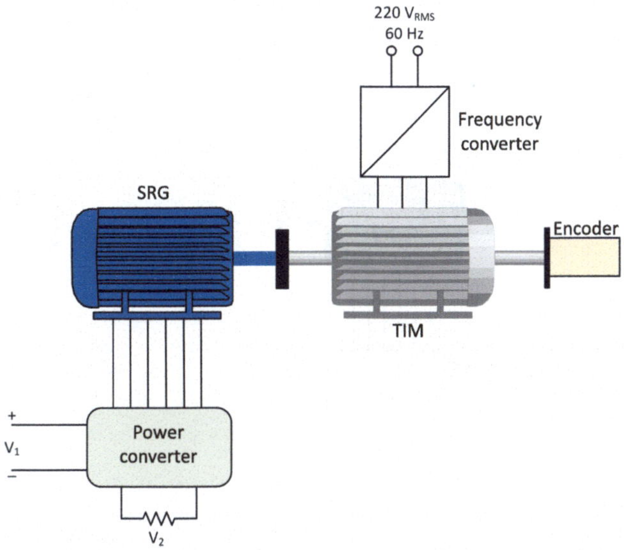

Fig. 5.3 Integration of the power converter and the switched reluctance generator

5.2.1 Configuration and Operation of the Power Converter

The power converter to be developed is the active hybrid bridge (AHB) converter. However, the goal is to construct a converter that facilitates the manipulation of the experimental platform while minimizing the need for hardware modifications for different types of test. To achieve this objective, the converter illustrated in Fig. 5.4 is being developed [3].

The AHB converter shown in Fig. 5.4 is integrated with three relays [4]. These electromechanical switches are incorporated into the converter circuit to enable reconfiguration to operate the switched reluctance machine as a motor or a generator, without requiring modifications to the experimental platform. The system

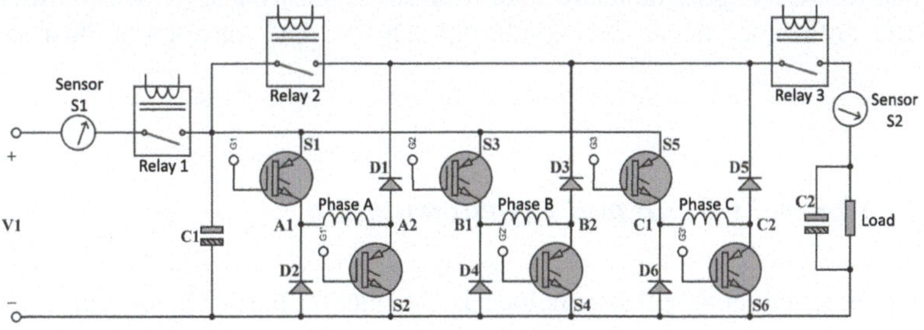

Fig. 5.4 Power converter for excitation of the switched reluctance generator

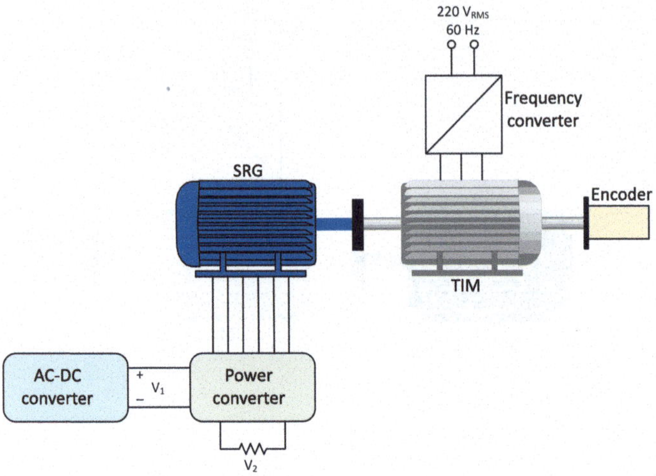

Fig. 5.5 Integration of the AC-DC converter into the experimental setup

controller automatically activates the relays to configure the converter for the desired operation. In Fig. 5.4, the devices labeled as Sensor $S1$ and Sensor $S2$ consist of voltage and current sensors, both using Hall-effect technology. These sensors measure the input voltage and current (excitation) of the generator, as well as the output voltage and current across the load, as such measurements are necessary to determine the efficiency of the SRG. The methodology is designed to implement power control of the switched resistance generator (SRG) using the excitation voltage V_1 as the variable manipulated. Consequently, a converter that allows for the application of variable DC voltage to the excitation bus of the AHB converter is necessary. This secondary circuit is a controlled AC-DC converter, forming the assembly illustrated in Fig. 5.5.

The AC-DC converter is connected to the three-phase network to enable a wide range of output DC voltage values, facilitating observation of the generator's behavior under different excitation voltage levels. Furthermore, the converter can apply voltage levels to the terminals of the Switched Reluctance Generator (SRG) that may drive the generator into core saturation, allowing for the assessment of the SRG's efficiency under such conditions. The AC-DC converter is illustrated in Fig. 5.6 [2].

5.2.2 *Control Circuit and Instrumentation*

The experimental setup will also include a controller for the activation and control of the switched resistance generator (SRG). This controller will be used for the proper switching of the AHB converter and the AC-DC converter for controlled excitation

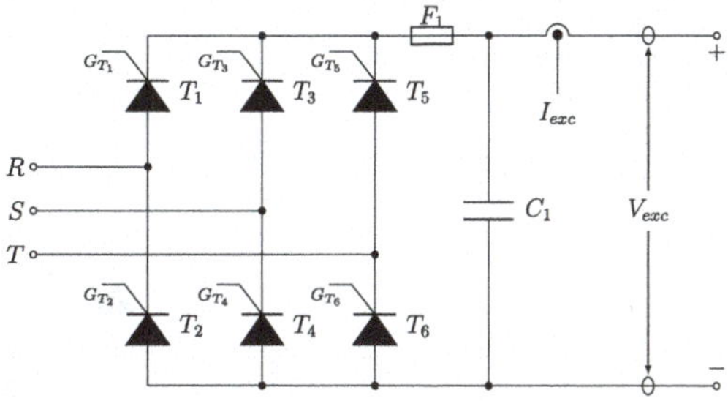

Fig. 5.6 Controlled AC-DC converter for producing excitation voltage in the windings of the switched reluctance machine

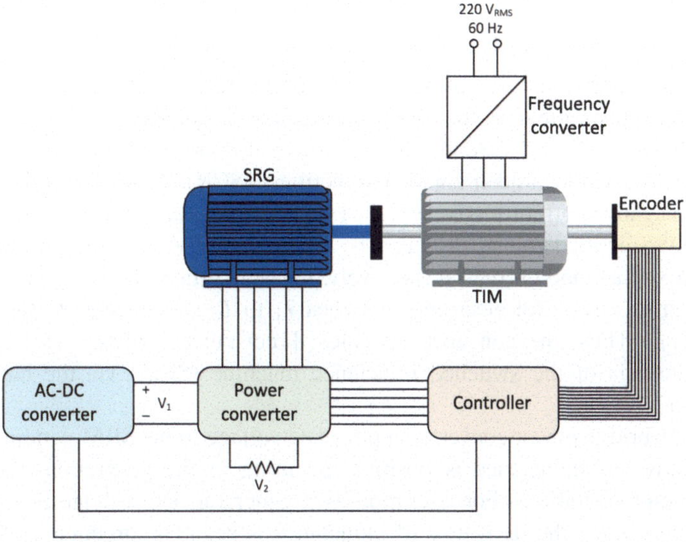

Fig. 5.7 Integration of the controller into the experimental setup

of the SRG. This assembly will enable the evaluation of the SRG's excitation behavior, facilitating studies on the integration of the generator with photovoltaic solar panels for excitation. Figure 5.7 illustrates the integration of the controller into the setup and provides an overview of the experimental platform [3].

5.3 Switched Reluctance Motor Operation

To activate and control the switched resistance machine (SRM), it is necessary to produce an adjustable direct current to modify the excitation voltage applied to the terminals L_1', L_2', L_3', and L_1'', L_2'', L_3''. In electrical drives, power converters with controlled rectifiers are used to obtain an adjustable direct current (DC) voltage from a constant alternating current (AC) voltage. Consequently, a fully controlled three-phase rectifier (FCTR) is designed to provide a wide range of excitation voltage

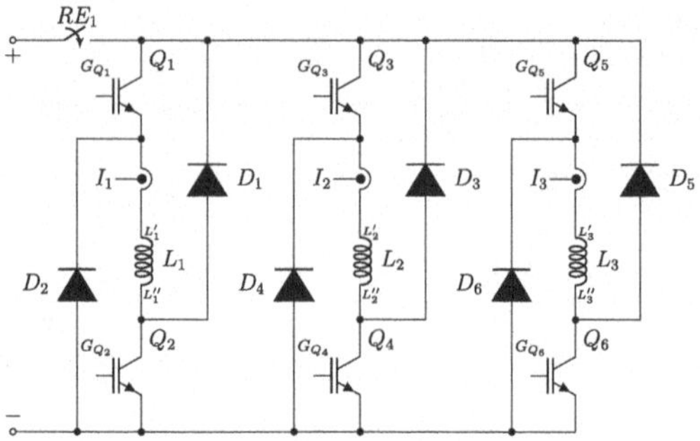

Fig. 5.8 Half-bridge converter for driving a three-phase switched reluctance motor, $\varsigma' = 3$

$Vexc$ for the operation of the SRM. Figure 5.6 illustrates the AC-DC converter and the respective instruments to measure the electrical parameters.

The phases R, S, and T represent the electrical network of the utility provider, while T_1 to T_6 denote silicon-controlled rectifiers (SCR). F_1 serves as the fuse for protection against current surges, and C_1 is the filtering capacitor that mitigates the ripples in the rectified voltage produced by the converter. V_{exc} and I_{exc} represent the excitation voltage and current, respectively, measured through Hall-effect sensors. The triggering circuit for electronic switches T_1 to T_6 is connected to the triggers G_{T_1} to G_{T_6}. Thus, the converter provides direct current voltage to be applied to the windings of the switched reluctance machine (SRM) via the half-bridge converter [2].

The half-bridge power converter applies DC voltage to the SRM windings when the derivative of inductance is positive, resulting in the generation of positive torque (motor operation). For operation as a generator, DC voltage is applied to the windings when the derivative of inductance is negative, producing a negative torque. Considering the three-phase switched reluctance machine with $\varsigma' = 3$, and based on Fig. 2.9, the half-bridge power converter to drive the three-phase SRM is presented in Fig. 5.8 [3].

The half-bridge converter features a triggering circuit connected to the gates G_{Q_1} to G_{Q_6}, with RE_1 representing the relay to connect the half-bridge converter in the FCTR power converter. The terminals of the switched reluctance motor windings, L_1', L_2', L_3', L_1'', L_2'', and L_3'', are connected to the half-bridge converter as shown in Fig. 5.8, while the Hall-effect current sensors are indicated by the reading signals I_1, I_2, and I_3. Electronic switches Q_1 to Q_6 are insulated gate bipolar transistors (IGBT), and diodes D_1 to D_6 are fast recovery silicon diodes. Figure 5.9 illustrates the SRM drive and control system, which consists of: (i) an FCTR AC-DC power converter, (ii) a half-bridge DC-DC power converter, (iii) the triggering circuit for the power converters, (iv) an absolute optical encoder for direct measurement of the rotor's instantaneous position, (v) instruments for measuring electrical voltages and currents using Hall-effect sensors, and (vi) a mechanical torque estimation system for the SRM.

The alternating current voltage V_{CA} is applied to the AC-DC converter (FCTR block in Fig. 5.9), thus producing the excitation voltage V_{exc} for the SRM windings

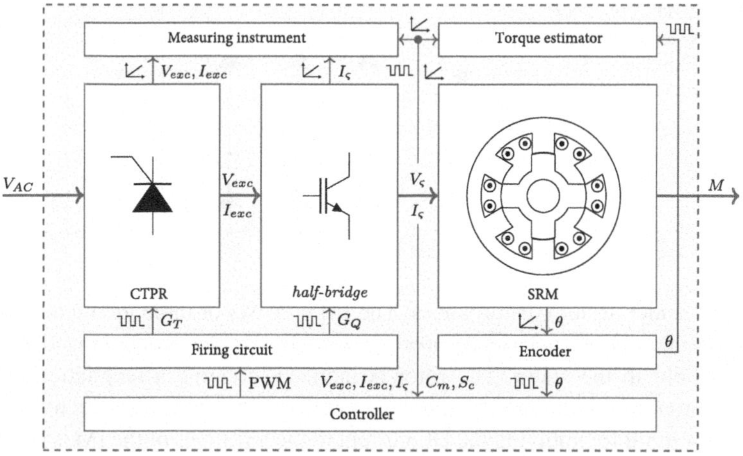

Fig. 5.9 Block diagram of the drive and control system for the switched reluctance motor

via the half-bridge converter. The half-bridge power converter generates voltage pulses V_ς for each SRM winding using the soft chopping switching technique. The encoder produces digital signals related to the instantaneous position of the rotor θ, which are received and interpreted by the system controller (controller block in Fig. 5.9). The controller applies pulse width modulation (PWM) signals to the triggering circuits of the AC-DC and half-bridge power converters (trigger circuit block in Fig. 5.9). For the operation of the AC-DC converter, the triggering circuit sends current pulses to the terminals G_{T_1} to G_{T_6} of the SCR switches, as shown in Fig. 5.6. This allows for the adjustment of the excitation voltage V_{exc} as desired. For the operation of the half-bridge converter, the triggering circuit applies voltage pulses to the terminals G_{Q_1} to G_{Q_6} of the IGBT switches, as illustrated in Fig. 5.8. Consequently, the excitation voltage V_{exc} is applied to the terminals L'_1, L'_2, L'_3, and L''_1, L''_2, L''_3 of each winding ς, according to the rotor position θ [2].

In Fig. 5.9, the measurement instruments consist of voltage and current sensors using Hall-effect technology, producing analog signals compatible with the voltage levels of the digital circuits present in the system. The mechanical torque estimator C_m is made up of a microcontroller that receives signals from sensors: (i) currents from the struts I_ς and (ii) rotor position θ. Thus, the mechanical torque C_m is calculated according to (2.7), and the result is transmitted as a PWM signal to the system controller. Synchronization and control signals S_c are exchanged between the torque estimator and the system controller. Electrical voltage information V_{exc}, excitation currents I_{exc}, I_ς, rotor speed ω, rotor position θ, mechanical torque C_m, among other data related to device processing and communication, are stored in the dataset represented by M.

The three-phase induction machine (IM) coupled to the shaft of the switched reluctance motor (SRM) operates as an electromagnetic brake, assisting in simulating different mechanical load behaviors and enabling analysis of drive and control

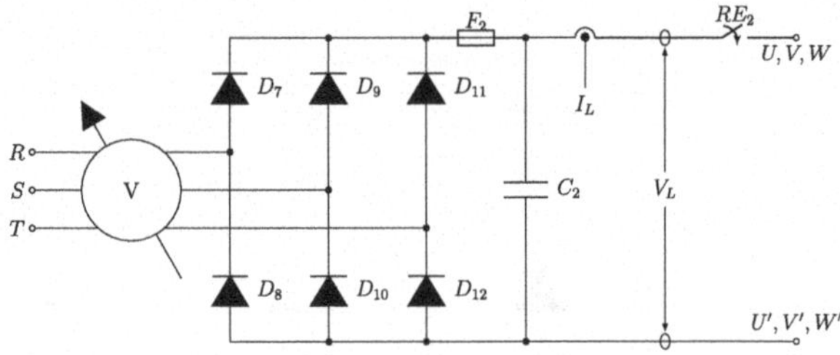

Fig. 5.10 AC-DC converter for generating a constant magnetic field in the three-phase induction machine

techniques under system disturbances. The IM consists of three identical windings connected in parallel, to which direct current is applied, generating a constant magnetic field in the stator. The rotor is induced, creating a magnetic field that opposes that of the stator, thereby allowing the IM to function as an electromagnetic brake. The circuit for applying the DC current to the terminals of the IM is illustrated in Fig. 5.10 [3].

In the case of Fig. 5.10, an uncontrolled three-phase rectifier (RTNC) is used as the AC-DC converter to supply constant current to the windings of the induction machine (IM). The DC current is regulated through an adjustable autotransformer connected to the electrical grid. The electromagnetic brake circuit consists of RE_2, representing the relay connecting the AC-DC converter to the IM windings, diodes D_7 to D_{12} as silicon rectifier diodes, fuse F_2 for protection against current surges, and capacitor C_2 to filter the ripple of rectified voltage from the converter. The measured voltage and current, denoted as V_L and I_L, respectively, are obtained using Hall-effect sensors.

5.4 Construction Characteristics

The constructed test setup is designed to allow the switched reluctance machine (SRM) to operate as a motor or a generator. Consequently, the three-phase induction machine (IM) with a squirrel cage rotor is coupled to the shaft of the SRM [2].

When the switched reluctance machine (SRM) operates as a generator, the induction machine (IM) functions as a motor; conversely, when the SRM operates as a motor, the IM acts as an electromagnetic brake. This coupling is designed for easy maintenance, allowing the SRM to be removed and replaced with another model if necessary. In this book, the SRM operates as a motor. The SRM was designed using the finite element method and was performed manually, with a focus on the context of this book. The electromechanical parameters are presented in Table 5.1 and are necessary for the computational model. Figure 5.11 illustrates the test setup constructed [3].

The AC-DC converter, specifically a fully controlled three-phase rectifier (FCTR), and the triggering circuit shown in Fig. 5.11 were constructed according to the electrical schematics described in Chap. 2. The FCTR serves as an adjustable

Table 5.1 Parameters of the switched reluctance motor used in this book

Parameter	Value
Number of phases	3
Number of poles in the stator	6
Number of poles in the rotor	4
Phase resistance [Ω]	3.25
Inertia $[(N \cdot m \cdot s^2)/rad]$	$5.897 \cdot 10^{-2}$
Viscous friction $[(N \cdot m \cdot s)/rad]$	$1.176 \cdot 10^{-3}$
Operating voltage [V]	180
Operating current [A]	3.20
Inductance with aligned rotor [mH]	255
Inductance with misaligned rotor [mH]	32

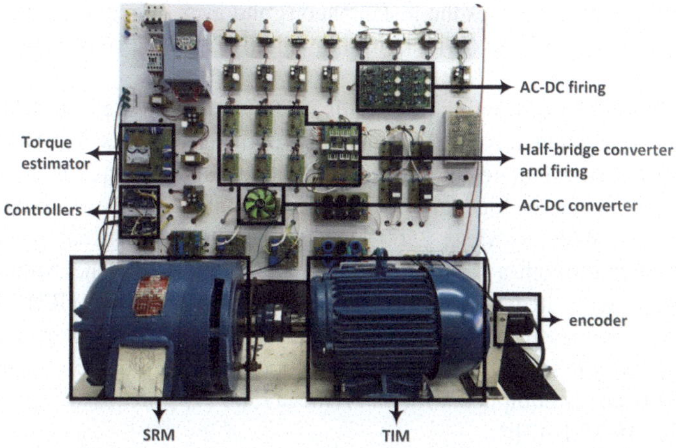

Fig. 5.11 Test setup for driving the switched reluctance machine

direct current source connected to the half-bridge converter. The electronic switches T_1 to T_6, illustrated in Fig. 5.6, are SK70DT08 SCRs [5] with operating voltage and current ratings of 800 V and 68 A, respectively. The capacitive filter C_1, also shown in Fig. 5.6, is an electrolytic capacitor with an equivalent capacitance of 2800 µF and can withstand voltages up to 1000 V [2].

The half-bridge converter shown in Fig. 5.11 was constructed according to operational and design specifications. The converter features electronic switches Q_1 to Q_6, which are IGBT IRG4PC50W [6] devices with operating voltage and current ratings of 600 V and 27 A, respectively, as illustrated in Fig. 5.8. The diodes D_1 to D_6, also illustrated in Fig. 5.8, are ultra-fast MUR1560 [7] diodes with operating voltage and current ratings of 600 V and 25 A, respectively. The trigger circuit for the IGBT IGR4PC50W switches of the half-bridge converter was designed accordingly. The drive and control setup for the switched reluctance motor (SRM), presented in Fig. 5.11, includes measurement instruments consisting of Hall-effect sensors for voltage and current measurement. The Hall voltage sensor is of type LV-

20P [8], capable of measuring up to 500 V, while the current Hall sensor is of type LA55P [9], capable of measuring up to 50 A.

These sensors provide an output signal that represents the instantaneous voltage proportional to the measured input voltage and current, respectively. The Hall sensor output signal is adjustable and connected to a second-order Butterworth filter to calculate the average value of the measured signal. The electrical schematics for the connections of the Hall sensors and the second-order Butterworth filter are not detailed. It is possible to measure the average and instantaneous excitation voltage, V_{exc} and I_{exc}, as well as the average and instantaneous currents in the SRM windings, I_1, I_2, and I_3. The torque estimator illustrated in Fig. 5.11 consists of a circuit built with an Arduino Due [10] prototyping board, which features an Atmel SAM3X8E ARM Cortex-M3 [11] processor operating at a frequency of 84 MHz. The torque estimator receives the instantaneous and average signals corresponding to the measurements of the currents in the windings and the rotor position of the SRM [12].

The measurement of rotation position is performed using the E6CP-AG5C [13] absolute encoder, which has a resolution of approximately 10 bits. The Due Arduino Prototyping Board calculates the instantaneous mechanical torque using (2.7) and sends the result as an analog signal to the system controller. The system controller consists of two Arduino Mega 2560 [14] prototyping boards that communicate with each other through a serial bus, as shown in Fig. 5.11. One Arduino Mega 2560 board is dedicated to the switching logic for driving the SRM, while the other focuses on measuring electromechanical variables, speed control, and communication with the supervisory system developed on the computer. Each Arduino Mega 2560 board is equipped with an ATmega2560 [15] processor operating at a frequency of 16 MHz [3].

The power converter to apply direct current to the electromagnetic brake is a non-controlled three-phase rectifier (RTNC), as illustrated in Fig. 5.10. This converter was constructed according to the electrical schematic. The RTNC serves as a source of direct current connected to the induction machine (IM), with the current adjusted through a variable autotransformer connected to the AC electrical grid. The electronic switches D_7 to D_{12} shown in Fig. 5.10 are of the 95PF [16] type, rated for an operating voltage of 800 V and a current of 95 A. The capacitive filter C_2, also illustrated in Fig. 5.10, is an electrolytic capacitor with an equivalent capacitance of 2800 μF, capable of withstand voltages up to 1000 V. This setup allows for the adjustment of the mechanical load applied to the shaft of the SRM, as demonstrated on the test bench for experimental trials, which highlights its main components in Fig. 5.7.

The power converter used to drive the induction machine (IM), which provides mechanical energy to the switched reluctance generator (SRG), is the WEG CFW500 [17] model. The induction motor has a power rating of 5 hp, and the model utilized is the W22 Plus, also from WEG. This setup allows for the application of torque to the shaft of the SRG, which is excited by a fully controlled three-phase rectifier, as illustrated in Fig. 5.6. The thyristor bridge employed consists of six SCR-type semiconductors encapsulated in a single housing. This device is

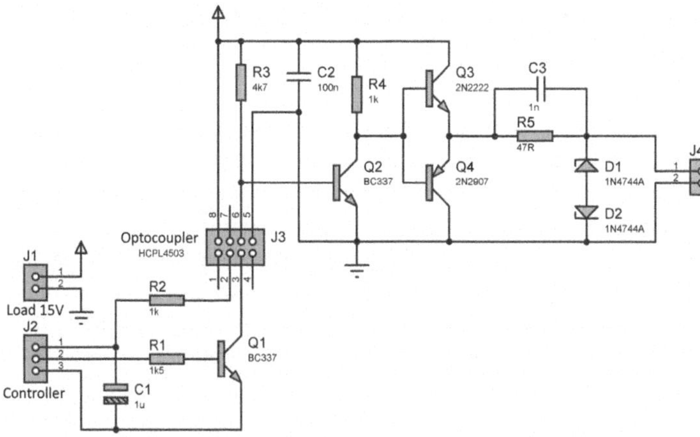

Fig. 5.12 Schematic of the driver circuit for integrating the controller with the IGBTs of the AHB converter

the SKDT115. manufactured by Semikron. A conditioning board was developed for this module, using the TCA785 [18] integrated circuit. The PWM signal from the controller is applied to a second-order Butterworth low-pass filter to recover only the DC component of the PWM signal, which is injected into the control terminal of the TCA785. This allows the AC-DC converter to receive continuous firing commands from 0 to 180 degrees, allowing the variation of the excitation voltage V_1 for the SRG [2].

Once the DC bus voltage from the AHB converter is obtained, the converter is constructed using IGBT-type transistors. The converter device used is the IRG4PC50W model from International Rectifier. The diodes utilized in the generation stage are MUR1560 models. The integration of the controller with the IGBTs of the power converter is accomplished through drivers to condition the control signals for these semiconductor switches. The circuit used for the six IGBTs of the AHB converter is illustrated in Fig. 5.12.

In Fig. 5.11, two sets of capacitors can be observed, corresponding to capacitors $C1$ and $C2$ in Fig. 5.4. Capacitors $C1$ and $C2$ consist of six 470 μF capacitors connected in parallel, resulting in a capacitance of 2.8 mF for both $C1$ and $C2$. The relays used to modify the structure of the AHB converter for operation as a motor and as a generator are the T92S7D12-24 [19] relays of TE Connectivity. The sensors used for monitoring the voltages and currents of the switched reluctance generator (SRG) and for the torque detection circuit are Hall-effect sensors. The LV25P sensors were used for voltage measurement, while the LA55P sensors were used for current measurement, both manufactured by LEM Semiconductors. Finally, the instantaneous position of the rotor θ is detected by an absolute encoder, Omron model E6CP-AG5C. This encoder has a resolution of 10 bits, allowing the controller to obtain angle values of approximately 0.35 degrees [3].

5.5 Experiments and Modeling of Inductance Computational Model

The mathematical and computational models of the switched reluctance machine (SRM) presented in the literature emphasize the importance of the inductance representation model. Since this book aims to implement optimization techniques for SRM control, it is desirable for the computational model to closely match the experimental platform (real system). Therefore, a methodology is required to obtain a model that accurately represents the inductances of the three windings of the SRG. To develop this model, inductance values are initially collected based on the rotor's angular position θ and the current i in the winding for one of the SRM windings [12].

5.5.1 Development of Parametric Regression and Experimental Testing

With the experimental inductance surface, consisting of a finite number of points, the next step is to develop the parametric regression algorithm to obtain the expression $L(\theta, i)$ that provides a simulation surface as close to the experimental surface values as possible. To find the expression that maps the simulation surface, an optimization process is used. In this process, the evaluation function is the difference between the values of the experimental surface and those of the surface formed by the mathematical expression. When the error between the two surfaces approaches zero, the mathematical expression accurately maps the behavior of the experimental surface. Thus, the goal is to obtain a continuous expression for the simulation surface with an infinite number of $L(\theta, i)$ points that can be used in the computational model shown in Fig. 2.20, which ultimately yields electrical and mechanical results for the SRG that closely match the experimental results [2].

To develop the simulator, it is necessary to identify the inductance profile model of the switched reluctance motor (SRM). In this case, an optimization process combined with a system identification technique known as parametric regression is used. Similarly to other nonlinear models that account for magnetic saturation, the proposed model relies on the experimental acquisition of inductance points for different angular positions θ and applied currents i, making it applicable to any SRM. Unlike the Fourier series approximation model, the approach used here derives the mathematical expression $L(\theta, i)$ that represents the inductances of the machine under study.

To determine the mathematical function corresponding to the inductance profile of the machine under study, it is necessary to conduct a locked-rotor test on the SRM. This test involves selecting one phase at random and using an AC source to align the stator and rotor poles of the chosen phase. The SRM rotor is then locked and the position $\theta = 0°$ is considered as the point of maximum

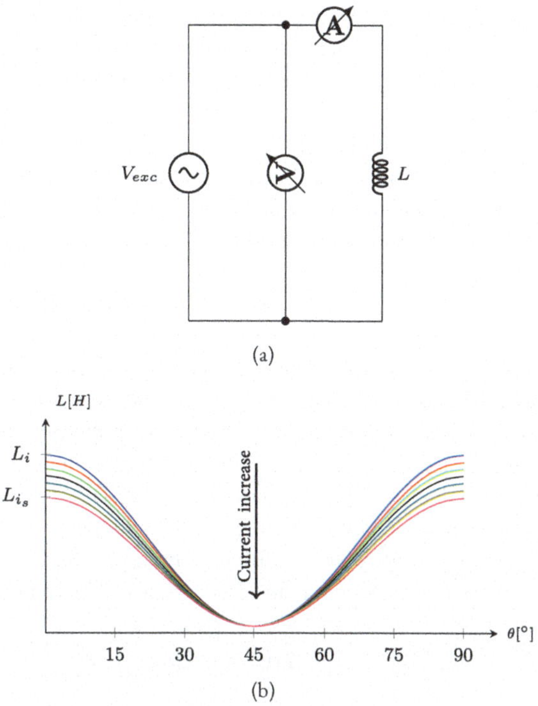

Fig. 5.13 Measurement for determining the inductance profile of the SRM: (**a**) electrical schematic for locked-rotor measurement and (**b**) theoretical inductance profile curves with increasing current

inductance. From this position, the rotor is incrementally locked at each step, with an adjustable AC source of frequency ϕ used to excite the winding. Figure 5.13a illustrates the electrical schematic used for data collection during the locked-rotor test. As an example, Fig. 5.13b shows inductance profile curves as the current value increases [3].

With the rotor locked at each position, the current is varied within the range $0 < i \leq i_s$, where i_s is the current value applied to the winding that causes magnetic flux saturation in the SRM, as shown in Fig. 2.16. From the experimentally collected data—voltage V, current i, electrical resistance R of the phase windings, and frequency of applied AC voltage ϕ—inductance can be calculated using the expression:

$$L = \left(\frac{V}{i}^2 - R^2\right)^{1/2} \cdot (2 \cdot \pi \cdot \phi)^{-1} \tag{5.1}$$

The expression (5.1) is a general formula for calculating the inductance of any inductor. This allows for the determination of phase inductance values L for each applied current i and rotor position θ, with a periodicity of 90° for the 6×4 SRM. Using the inductance values obtained from (5.1), parametric regression is used to derive the mathematical expression $L_m(\theta, i)$ that represents the inductance profile. Unlike other methods that model inductance profiles, the proposed method analytically expresses the inductance profile through parametric regression. The parametric regression method is designed for use in white-box, gray-box, or black-box systems and requires input variable values, system output values, and the mathematical model [20]. In this specific application, the system is a gray-box model, as defined by the mathematical model. To find the surface $L_m(\theta, i)$ that captures the characteristics of the measured inductance in the machine under study $L_s(\theta, i)$, the following is used:

$$L_m(\theta, i) = p_u + (f_1 \cdot f_2 \cdot f_3) \tag{5.2}$$

where f_1 is the expression defined by the Gaussian distribution, which exhibits a parabolic behavior near the origin and smooth continuity beyond the origin; f_2 is the expression defined by Legendre polynomials, suitable for fitting data from continuous functions; and f_3 is the expression defined by the Fourier trigonometric series, which exhibits periodic behavior. These three models are combined, as they represent the potential behaviors of inductance surfaces [12]. Each expression f_1, f_2, and f_3 represents a different behavior that can be modified by adjusting the coefficients, exponents, and frequencies, as given by:

$$f_1(\theta, i) = p_{(u+1)} \cdot e^{\left[p_{(u+2)} \cdot i + p_{(u+3)} \cdot \theta\right]} \tag{5.3}$$

$$\begin{aligned} f_2(\theta, i) = & p_{(u+4)} \cdot i^{p_{(u+5)}} + p_{(u+6)} \cdot \\ & \theta^{p_{(u+7)}} + \cdots + p_{(v-3)} \cdot i^{p_{(v-2)}} + p_{(v-1)} \cdot \theta^{p_{(v)}} \end{aligned} \tag{5.4}$$

$$\begin{aligned} f_3(\theta, i) = & p_w \cdot \sin\left[p_{(w+1)} \cdot i + p_{(w+2)}\right] + \\ & p_{(w+3)} \cdot \sin\left[p_{(w+4)} \cdot \theta + p_{(w+5)}\right] + \cdots + \\ & p_{(j-5)} \cdot \sin\left[p_{(j-4)} \cdot i + p_{(j-3)}\right] + \\ & p_{(j-2)} \cdot \sin\left[p_{(j-1)} \cdot \theta + p_{(j)}\right] \end{aligned} \tag{5.5}$$

where p_u to $p_{(u+3)}$ are parameters related to the Gaussian distribution, $p_{(u+4)}$ to p_v are parameters associated with Legendre polynomials, with $1 \leq u \leq v$, and p_w to p_j are the parameters related to the trigonometric series, with $1 \leq w \leq j$. The total number of parameters to be optimized, $j + v$, includes coefficients, exponents, and frequencies that define the characteristics of $L_m(\theta, i)$. These parameters are unknown and are determined using an optimization method. The inductance data obtained from the experimental measurements through (5.1) is organized such that L is represented in the matrix $L_s(\theta, i)$ with dimension a_1, comprising the measured values of θ and i. The evaluation function of the optimization method employs the expressions (5.1) and (5.2) to compare the obtained results [3].

5.5.2 *Optimization Algorithm and Model Evaluation*

Figure 4.3 illustrates the flow diagram of the parametric regression algorithm, where $n_p = j + v$. The matrices containing the system output data $S_s = L_s(\theta, i)$ and the model output data $S_m = L_m(\theta, i)$ have different dimensions, a_1 and a_2, respectively, and need to be normalized. For this specific case, where $S_{s_{(a_1)}}$ has size a_1 and $S_{m_{(a_2)}}$ has size a_2, normalization blocks should be added below the system and model blocks in Fig. 4.3. This ensures that S_s and S_m have the same dimensions and can be compared in the comparison block shown in Fig. 4.3 [2].

The optimization process is performed using a hybrid optimization algorithm due to the highly nonlinear nature of the system. The hybrid algorithm used combines the deterministic quasi-Newton method (QNM) with the heuristic Genetic Algorithm (GA). The evaluation function used in the optimization process represents the difference between various regions of the surfaces $L_m(\theta, i)$ and $L_s(\theta, i)$ and is given by:

$$f_r(p_1.p_2.\cdots, p_{n_p}) = \frac{1}{4}\cdot\left(f_{a_r} + f_{a_r}\Big|_{\theta=0^\circ} + f_{a_r}\Big|_{\theta=90^\circ} + f_{a_r}\Big|_{\theta_{\min}\leq\theta\leq\theta_{\max}}\right) \tag{5.6}$$

where:

$$f_{a_r}(p_1.p_2.\cdots, p_{n_p}) = \int_0^i\int_0^\theta \left|\frac{L_s(\theta, i) - L_m(\theta, i)}{L_s(\theta, i)}\right| d\theta di \tag{5.7}$$

where f_{a_r} represents the difference between the surfaces $L_m(\theta, i)$ and $L_s(\theta, i)$. Certain regions of the surface $L_m(\theta, i)$ were prioritized to achieve better adjustments, as these regions are more significant in characterizing the inductance of the SRM. Thus, the regions (i) $\theta = 0^\circ$, (ii) $\theta = 90^\circ$, and (iii) $\theta_{\min} \leq \theta \leq \theta_{\max}$, defined in (5.6), are assigned higher weights in the optimization process [12].

5.6 Elaboration and Validation of the Induction Surface

The next step involves obtaining the set of points $L(\theta, i)$ to determine the inductance surface of the switched resistance generator (SRG). To implement optimization techniques in SRG control, the computational model must accurately represent the

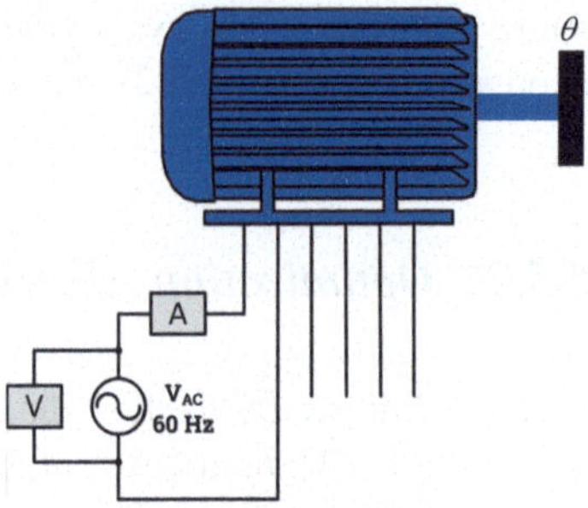

Fig. 5.14 Circuit used to collect inductance points

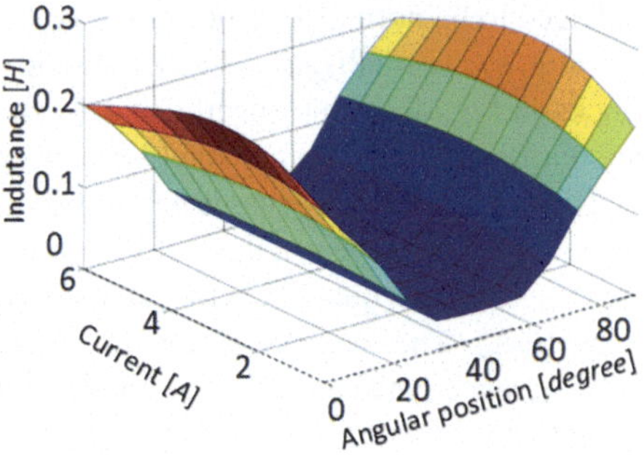

Fig. 5.15 Experimental inductance surface of the switched reluctance generator

real system. Therefore, a methodology that enables the development of a model that enables the inductances of the three windings of the SRG to be highly precise is required. Initially, inductance values are collected as functions of the angular position of the rotor, θ, and the current, i, applied to one of the windings. Figure 5.14 illustrates the circuit used to obtain these $L(\theta, i)$ values [3].

A sinusoidal voltage is applied to the terminals of one SRG winding. From the measured voltage, current, and frequency of the applied voltage, the required values are collected. This procedure is repeated for various values of θ to construct the inductance surface. The experiment is carried out on only one of the three windings of the SRG, as the other windings share identical construction, differing only by a mechanical displacement of 60 degrees relative to each other [12].

The diagram in Fig. 5.14 allows the collection of 108 inductance points. These points correspond to measurements at the angular positions of 0°, 6°, 11°, 16°, 22°, 27°, 33°, 39°, and 45°, resulting in nine points spanning from the aligned position (0°) to the fully unaligned position (45°). For each angular position, the current applied to the pinion is varied from 0 A to 6 A in increments of 0.5 A. This dataset facilitates the construction of the experimental inductance surface, as shown in Fig. 5.15 [2].

This book aims to derive an expression that provides a continuous inductance surface, that is, it contains an infinite number of points $L(\theta, i)$. Such a continuous surface can positively contribute to the computational model by enhancing both the accuracy of the results compared to the experimental platform and the simulation time. Consequently, an expression $L(\theta, i)$ has been developed with three components: (i) exponential distribution f_1, (ii) Taylor series f_2, and (iii) Fourier series f_3, as indicated in (5.8).

$$L_{CONT}(\theta, i) = p_1 + (f_1 \cdot f_2 \cdot f_3) \tag{5.8}$$

The terms f_1, f_2, and f_3 are defined as:

$$f_1(\theta, i) = p_2 \cdot e^{p_3} \cdot e^{p_4} \tag{5.9}$$

$$f_2(\theta, i) = p_5 \cdot i^{p_6} + p_7 \cdot \theta^{p_8} + p_9 \cdot i^{p_{10}} + p_{11} \cdot \theta^{p_{12}} + p_{13} \cdot i^{p_{14}} + p_{15} \cdot \theta^{p_{16}} \tag{5.10}$$

$$\begin{aligned} f_3(\theta, i) = {} & p_{17} \cdot \sin(p_{18} \cdot i + p_{19}) + p_{20} \cdot \sin(p_{21} \cdot \theta + p_{22}) \\ & + p_{23} \cdot \sin(p_{24} \cdot i + p_{25}) + p_{26} \cdot \sin(p_{27} \cdot \theta + p_{28}) \\ & + p_{29} \cdot \sin(p_{30} \cdot i + p_{31}) + p_{32} \cdot \sin(p_{33} \cdot \theta + p_{34}) \end{aligned} \tag{5.11}$$

The expression (5.8) contains 34 parameters that need to be determined in order to obtain a continuous inductance surface that closely matches the experimental surface shown in Fig. 5.15. To derive the parameters for the expressions (5.8), (5.9), (5.10), and (5.11), a technique known as parametric regression was used, in which the parameters of a function are obtained through an optimization process. In this case, a hybrid genetic algorithm was applied to approximate the continuous surface with the experimental surface. Thus, the evaluation function of this algorithm corresponds to the difference between the experimental surface L_{EXP} and the continuous surface L_{CONT}. Therefore, the objective of the algorithm is to minimize this difference between the surfaces, as indicated in (5.12) [3].

$$f_a(p) = \frac{\sum |L_{EXP}(\theta, i) - L_{CONT}(\theta, i)|}{\sum L_{EXP}(\theta, i)} \tag{5.12}$$

With the continuous inductance surface of the switched reluctance generator (SRG) established, the computational model can be evaluated using accuracy as an indicator by comparing it with the results obtained experimentally, both for the electrical quantities of the SRG and for the mechanical quantities [12]. The following section presents the simulation results using the continuous inductance surface.

References

1. R. Krishnan et al., High reliability SRM drive system for aerospace applications, in *IECON'03. 29th Annual Conference of the IEEE Industrial Electronics Society (IEEE Cat. No.03CH37468)* (2003). https://doi.org/10.1109/IECON.2003.1280202

2. M.R. Cunha Reis et al., Optimized techniques for driving and control of the switched reluctance motor to improve efficiency. Control Eng. Pract. **90**, 1–18 (2019)
3. W.R.H. Araujo et al., Efficiency enhancement of switched reluctance generator employing optimized control associated with tracking technique. Energies **14**(24), 8388 (2021). https://doi.org/10.3390/en14248388
4. K. Ogata et al., *Modern Control Engineering* (Prentice Hall India, Indore, 2009)
5. Semikron International GmbH, *SK70DT08 Datasheet.* Accessed on: January 29, 2025. Semikron (2025). https://datasheet.octopart.com/SK70DT08-SemiKron-datasheet-10903793.pdf
6. Infineon Technologies AG, *IRG4PC50W Datasheet.* Accessed on: January 29, 2025. Infineon Technologies (2004). https://www.infineon.com/dgdl/Infineon-IRG4PC50W-DataSheet-v01_00-EN.pdf?fileId=5546d462533600a401535647cdad22fb
7. ON Semiconductor, *MUR1560 Datasheet.* Accessed on: January 29, 2025. ON Semiconductor (2014). https://www.onsemi.com/download/data-sheet/pdf/mur1520-d.pdf
8. LEM International SA, *LV 20-P Datasheet.* Accessed on: January 29, 2025. LEM (2004). https://media.digikey.com/pdf/data%20sheets/lem%20usa%20pdfs/lv%2020-p.pdf
9. LEM International SA, *LA 55-P Datasheet.* Accessed on: January 29, 2025. LEM (2023). https://www.lem.com/sites/default/files/products_datasheets/la_55-p_v19.pdf
10. Arduino LLC, *Arduino Due Product Reference Manual.* Accessed on: January 29, 2025. Arduino (2025). https://docs.arduino.cc/resources/datasheets/A000062-datasheet.pdf
11. Atmel Corporation, *SAM3X/SAM3A Series Datasheet.* Accessed on: January 29, 2025. Atmel (2012). https://ww1.microchip.com/downloads/en/DeviceDoc/Atmel-11057-32-bit-Cortex-M3-Microcontroller-SAM3X-SAM3A_Datasheet.pdf
12. L.D.S. Morais et al., Enhancement of the performance of switched reluctance generators in low wind speed conditions using advanced tracking techniques. Int. J. Energy Res. **2024**(1), 5541150 (2024)
13. Omron Corporation, *E6CP-AG5C Absolute Encoder Datasheet.* Accessed on: January 29, 2025. Omron (2017). https://assets.omron.eu/downloads/latest/datasheet/en/e6cp-a_datasheet_en.pdf
14. Arduino LLC, *Arduino Mega 2560 Rev3 Product Reference Manual.* Accessed on: January 29, 2025. Arduino (2025). https://docs.arduino.cc/resources/datasheets/A000067-datasheet.pdf
15. Microchip Technology Inc., *ATmega640/1280/1281/2560/2561 Datasheet.* Accessed on: January 29, 2025. Microchip Technology (2014). https://ww1.microchip.com/downloads/en/DeviceDoc/Atmel-2549-8-bit-AVR-Microcontroller-ATmega640-1280-1281-2560-2561_Datasheet.pdf
16. Vishay Semiconductors, *VS-95PF(R) (W) Series Standard Recovery Diodes Datasheet.* Accessed on: January 29, 2025. Vishay (2018). https://www.vishay.com/docs/93532/vs-95pfrseries.pdf
17. WEG Electric Equipment S.A., *CFW500 - Frequency Inverter.* Accessed on: January 29, 2025. WEG (2024). https://static.weg.net/medias/downloadcenter/h3d/h0f/WEG-CFW500-50036259-en.pdf
18. Infineon Technologies AG, *TCA785 - Phase Control IC Datasheet.* Accessed on: January 29, 2025. Infineon (2002). https://www.farnell.com/datasheets/1836360.pdf
19. TE Connectivity, *T92S7D12-24 Relay Datasheet.* Accessed on: January 29, 2025. TE Connectivity (2025). https://www.te.com/en/product-1-1393212-0.html
20. F.A. Gomes et al., Heuristic regression method for descriptive data analysis, in *2016 IEEE 16th International Conference on Environmentand Electrical Engineering (EEEIC)* (IEEE, 2016), pp. 1–5

Chapter 6
Optimization and Validation of Switched Reluctance Generator Models

6.1 Computational Model

For the simulation, the computational model receives initial values related to the machine's characteristics, as illustrated in Fig. 2.20. These parameters are defined based on the characteristics of the generator, as represented in Fig. 5.12. Table 6.1 presents the characteristics of the SRG used in this book [1].

As described in Chap. 2, although Table 6.1 presents only a single value for L_{MAX} and L_{MIN}, the inductance of the SRG windings depends not only on the angular position but also on the current. Thus, Table 6.1 shows only the maximum and minimum inductance values obtained experimentally using the system illustrated in Fig. 5.15. The initial verification of the results of the computational model was performed by applying an excitation voltage V_1 of 150 V to the SRG, with a rotor speed of 1000 rpm. The load connected to the generator output was resistive, with a value of 32 Ω. The windings were kept constant, with the turn-on angle θ_{ON} set at 0°, corresponding to the point of maximum inductance (aligned position) and the turn-off angle θ_{OFF} set at 30°. Figure 6.1 illustrates the inductance in one of the SRG windings and the logical switching command associated with it [2].

Figure 6.2 presents the simulation results for voltage and current in one of the SRG windings. It can be observed that, at the end of the magnetization (excitation) phase, when the switching reaches the value of θ_{OFF}, the current in the winding begins to decrease but maintains the same direction. During this stage, energy generation occurs and the current is directed to the load through the diodes of the AHB converter. Figure 6.3 shows the simulation results for the input and output powers of the SRG.

In Fig. 6.3, the blue curve represents the input electrical power (P_1) used for the excitation of the generator, which has a steady-state value of 604.2 W. The red curve corresponds to the generated electrical power (P_2), with the SRG producing 1290.3W under the conditions described above steady state. Finally, the black curve indicates the mechanical power applied to the shaft of the SRG (P_{MEC}), which

W. P. Calixto et al., *Control and Tracking Techniques for Switched Reluctance Machines*, Power Systems, https://doi.org/10.1007/978-3-031-86727-9_6

Table 6.1 Characteristics of the simulated machine

Parameter	Value
Minimum phase inductance L_{MIN}	31.95 mH
Maximum phase inductance L_{MAX}	255.05 mH
Viscous friction D	$1 \cdot 10^{-4}$ kg $\cdot$ m^2/s
Inertia J	$8 \cdot 10^{-3}$ kg $\cdot$ m^2
R_E	3.25 Ω
Stator poles	6
Rotor poles	4

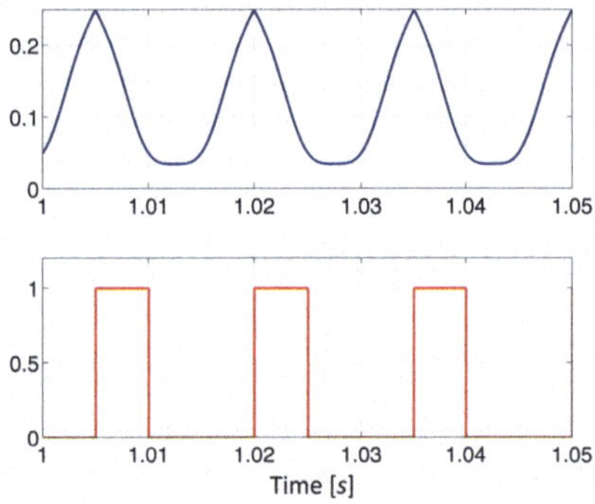

Fig. 6.1 Inductance and logical switching command

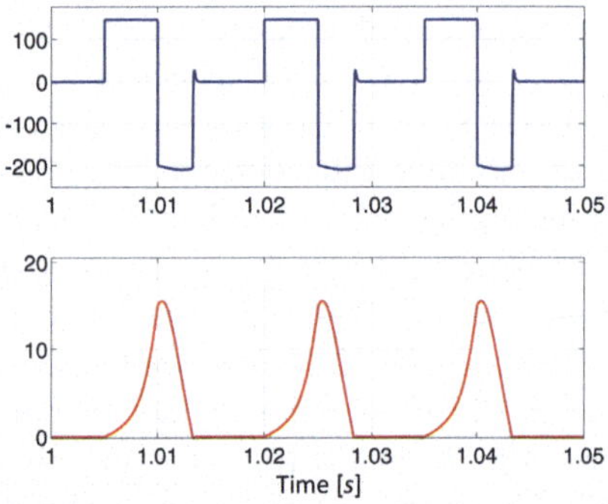

Fig. 6.2 Voltage and current in one of the SRG windings

reaches a steady-state value of 972 W. These power values result in a total efficiency of 81.8% for the SRG. As previously described, this book aims to optimize the turn-off angle θ_{OFF} to enhance efficiency, regardless of the excitation voltage applied by the power controller. Figure 6.4 graphically illustrates the total efficiency (n_T) of the SRG obtained through the computational model.

To verify the reliability of the values obtained in the simulation, they are compared with those obtained experimentally on the test bench. Initially, only electrical quantities are analyzed, such as the excitation current I_1 and the voltage generated V_2. In addition, the results of the computational model are compared with those of other inductance representation models. Each of these quantities is evaluated using the values provided by the computational model based on the following inductance representation models: (i) linear model, (ii) nonlinear model

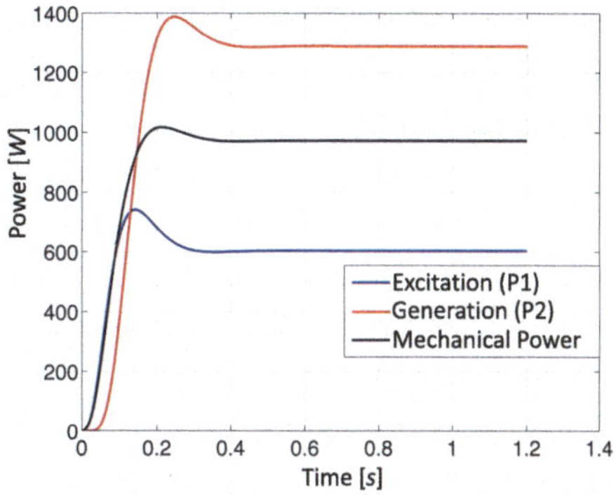

Fig. 6.3 Input and output powers of the SRG

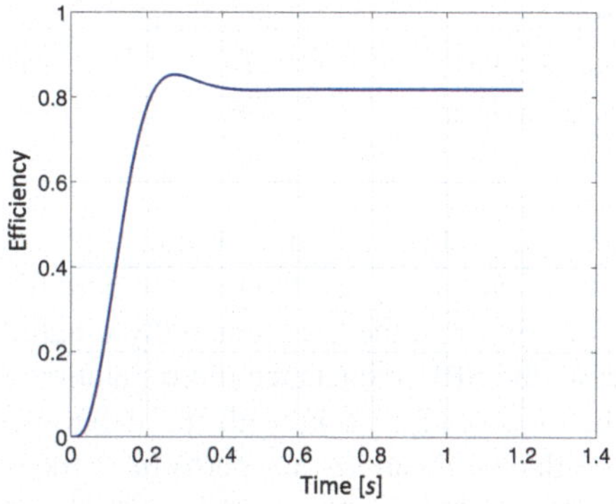

Fig. 6.4 Total efficiency of the SRG

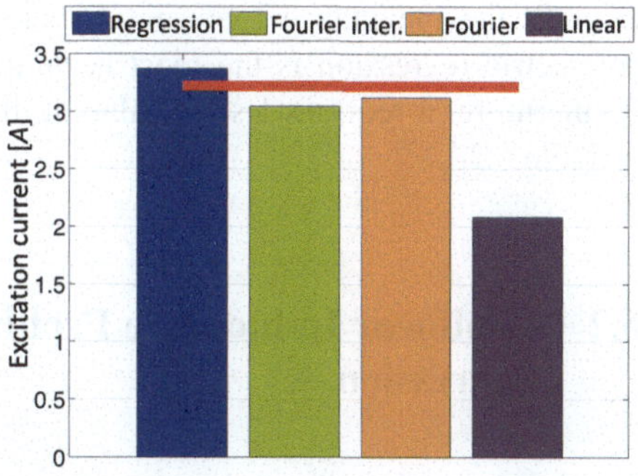

Fig. 6.5 Excitation current I_1 compared between computational and experimental models

using Fourier series, (iii) nonlinear model using Fourier series with polynomial interpolation between inductance points, and (iv) inductance representation model using parametric regression, as presented in this section. Figure 6.5 shows the excitation current values of the SRG for the different models of inductance representation compared to the value obtained from the experimental platform [3].

In Fig. 6.5, the red horizontal line represents the experimental value obtained for the SRG excitation current, which is 3.24 A in steady state. The computational models using Fourier series and parametric regression for inductance representation are observed to demonstrate a closer alignment with the experimental model,

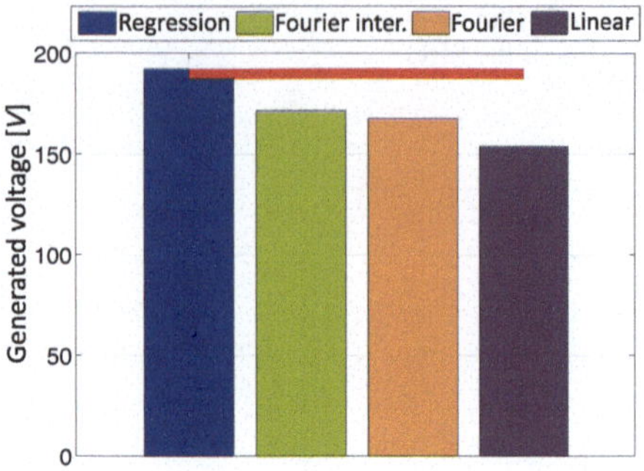

Fig. 6.6 Generated voltage V_2 compared between computational and experimental models

Table 6.2 Average deviation of computational models and simulation times

Parameters	Regression	Fourier (INT)	Fourier	Linear
Deviation [%]	3.85	10.96	13.68	31.02
Simulation time [s]	55.16	287.17	64.38	17.99

yielding values of 3.04 A and 3.37 A, respectively. Figure 6.6 presents the corresponding comparison for the output voltage of the SRG [3].

In Fig. 6.6, the experimentally obtained value is 190 V in steady state. In this case, the SRG inductance representation model using parametric regression shows the closest agreement with the experimental model, with a deviation of 1.08%. Similar verifications are performed for other excitation voltage values and rotor speeds, subjecting the computational model to a wide range of test conditions. Overall, the inductance representation model for the SRG windings employing parametric regression is the most accurate compared to the experimental model. Furthermore, it requires less simulation time between nonlinear models, as shown in Table 6.2.

6.2 Nonlinear Inductance Profile Using Parametric Regression

For the simulation, the expression that best fits the inductance surface profile of the studied SRM is used. To obtain this expression, bench tests are performed by applying an AC voltage to the phase A winding. The machine's rotor remains locked, and the AC voltage is gradually increased to modify the current until the onset of magnetic flux saturation is observed. The current ranges from 0.5 A to 6 A, while the rotor is positioned from 0° to 90° in increments of 3°. The AC voltage and current values are applied to Eq. (5.1), with $\phi = 60$ Hz and $R = 3.25\ \Omega$. The experimental inductance values $L_s(\theta, i)$ are shown on the blue surface in Fig. 6.7a [2].

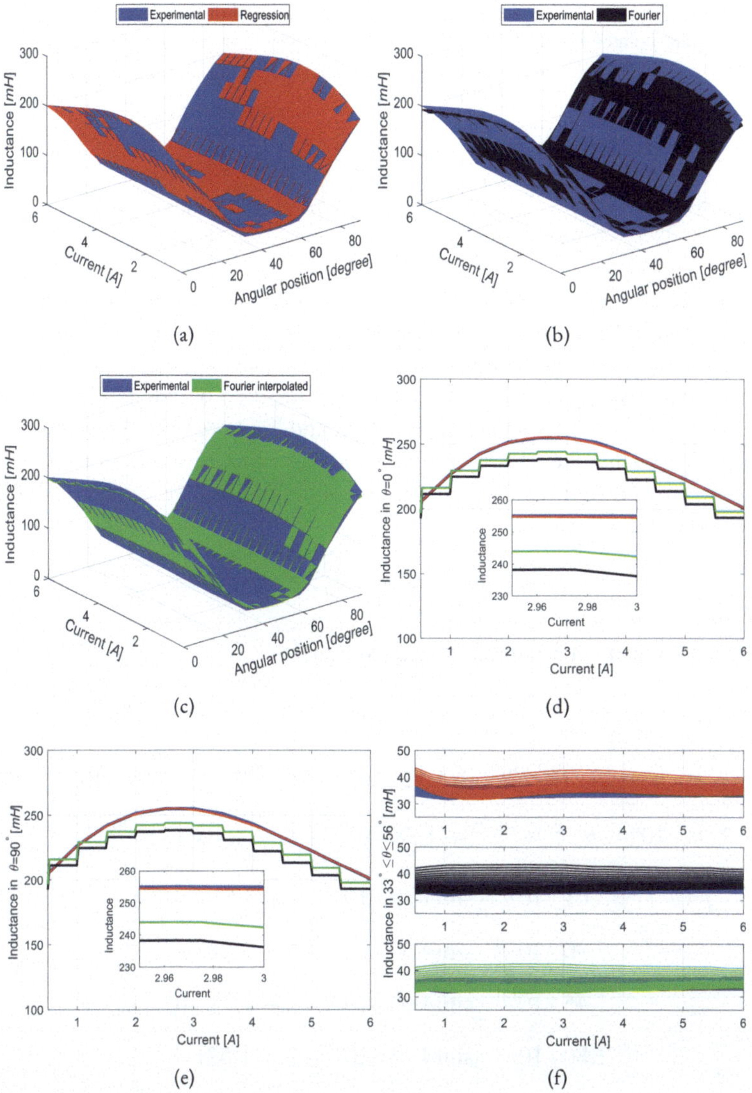

Fig. 6.7 Inductance profile: (**a**) experimental inductance surface and parametric regression, (**b**) experimental inductance surface and Fourier, (**c**) experimental inductance surface and interpolated Fourier, (**d**) inductances at $\theta = 0°$, (**e**) inductances at $\theta = 90°$, and (**f**) inductances in the range $33° \leq \theta \leq 56°$

Based on the experimental inductance data $L_s(\theta, i)$ obtained from the test and the Eq. (5.1), the parametric regression algorithm is executed for the values of $0° \leq \theta \leq 90°$ and $0.5\ \text{A} \leq i \leq 6\ \text{A}$ to determine the parameters p in the expression (5.2). The hybrid optimization algorithm, which combines a genetic algorithm

Table 6.3 Comparison among inductance profile calculation methods: parametric regression, Fourier, and interpolated Fourier

Simulation	(5.6) [%]	(5.7) [%] $\forall\theta$	$\theta = 0°$	$\theta = 90°$	$33° \leq \theta \leq 56°$	T_s [s]
Regression	0.117	0.301	0.023	0.027	0.116	0.030
Fourier	0.691	0.392	1.138	1.138	0.097	0.272
Interpolated	0.490	0.422	0.710	0.710	0.118	33.730

(GA) and the quasi-Newton method, is configured with an initial population of 20 individuals, a maximum generation limit of $g_{\max} = 100$, uniform mutation, tournament selection, and heuristic crossover. The evaluation function f_r, defined in (5.6), considers $\theta_{\min} = 33°$ and $\theta_{\max} = 56°$, justified by the nonlinearity of the region $33° \leq \theta \leq 56°$, as shown in Fig. 6.7a and Table 6.3 [2]. The parameters p are determined using the hybrid optimization algorithm, resulting in the optimized analytical function $L_m(\theta, i)$, which represents the inductance surface of the SRM studied and is expressed as:

$$L(\theta, i) = 1.10 + (f_1 \cdot f_2 \cdot f_3) \tag{6.1}$$

where f_1, f_2, and f_3 are defined by (6.2) to (6.4).

$$f_1(\theta, i) = 3.41 \cdot 10^{-2} \cdot e^{[(-2.71 \cdot 10^{-1} \cdot i) - (2.76 \cdot 10^{-4} \cdot \theta)]} \tag{6.2}$$

$$\begin{aligned} f_2(\theta, i) &= 6.29 \cdot i^{9.22 \cdot 10^{-1}} + 1.52 \cdot 10^{-5} \cdot \theta^{(8.39 \cdot 10^{-4})} - 3.91 \cdot 10^{-1} \cdot i^{0.48} \\ &+ 2.03 \cdot 10^{-1} \cdot \theta^{(-5.97 \cdot 10^{-6})} + 1.88 \cdot 10^{-2} \cdot \theta^{8.97} - 0.84 \cdot i^{5.30 \cdot 10^{-1}} \end{aligned} \tag{6.3}$$

$$\begin{aligned} f_3(\theta, i) =& 3.01 \cdot 10^{-1} \cdot \sin(1.42 \cdot 10^{-1} \cdot i + 5.51 \cdot 10^{-1}) + \\ & 7.90 \cdot 10^{-1} \cdot \sin(4.37 \cdot 10^{-2} \cdot \theta + 2.48) + \\ & 1.45 \cdot 10^{-1} \cdot \sin(4.92 \cdot 10^{-1} \cdot i + 4.95 \cdot 10^{-8}) - \\ & 1.80 \cdot 10^{-1} \cdot \sin(4.23 \cdot 10^{-2} \cdot \theta + 1.53) - \\ & 4.70 \cdot 10^{-2} \cdot \sin(4.81 \cdot 10^{-1} \cdot i - 1.50 \cdot 10^{-1}) + \\ & 4.32 \cdot 10^{-2} \cdot \sin(1.72 \cdot 10^{-1} \cdot \theta + 5.96) \end{aligned} \tag{6.4}$$

The derivatives of inductance with respect to θ are given by:

$$\begin{aligned}\frac{\partial L(\theta,i)}{\partial \theta} =& f_1 \cdot \{[3.41 \cdot 10^{-2} \cdot \cos(4.30 \cdot 10^{-2} \cdot \theta + 2.48) - \\ & 7.95 \cdot 10^{-3} \cdot \cos(4.22 \cdot 10^{-2} \cdot \theta + 1.53) + \\ & 7.68 \cdot 10^{-3} \cdot \cos(1.71 \cdot 10^{-1} \cdot \theta + 5.96)] \cdot f_2 - \\ & f_1 \cdot [1.60 \cdot 10^{-1} \cdot \theta^{-9.97} - 1.27 \cdot 10^{-8} \cdot \theta^{-9.90 \cdot 10^{-1}} + \\ & 1.23 \cdot 10^{-6} \cdot \theta^{-1}] \cdot f_3 - 9.43 \cdot 10^{-6} \cdot \\ & e^{[(-2.71 \cdot 10^{-1} \cdot i) - (2.76 \cdot 10^{-1} \cdot \theta)]} \cdot f_2 \cdot f_3\}\end{aligned} \tag{6.5}$$

The optimizer identifies the evaluation function, defined by (5.6), with a value of $f_r = 0.117\%$ for the parametric regression method, ensuring optimized results compared to those obtained with other methods. The remaining methods are evaluated using the same evaluation function in (5.6), resulting in 0.691% for the Fourier method and 0.490% for the interpolated Fourier method [1]. Figure 6.7 shows the inductances obtained experimentally and by the parametric regression, Fourier, and interpolated Fourier methods. In the region corresponding to all θ and i, considered in the surface difference function given by (5.7), the obtained values are $f_{a_r} = 0.301\%$ for the parametric regression method (Fig. 6.7a), 0.392% for the Fourier method (Fig. 6.7b), and 0.422% for the interpolated Fourier method (Fig. 6.7c).

In the region $\theta = 0°$, the evaluation for the parametric regression method is $f_{a_r} = 0.023\%$, while it is 1.138% for the Fourier method and 0.710% for the interpolated Fourier method, as illustrated in Fig. 6.7d. In the region $\theta = 90°$, the evaluation values are $f_{a_r} = 0.027\%$ for the parametric regression method, 1.138% for the Fourier method and 0.710% for the interpolated Fourier method [4], as shown in Fig. 6.7e.

6.3 Design and Construction of the Indirect Torque Detection System

In the mathematical modeling described in (2.22) and (2.23), the electromagnetic torque of the switched reluctance machine is defined as the rate of change of the coenergy. Thus, (2.23) can be rewritten for the three phases of the SRG as:

$$T_{EM} = \frac{1}{2} \cdot i^2 \cdot \frac{\partial L_A}{\partial \theta} + \frac{1}{2} \cdot i^2 \cdot \frac{\partial L_B}{\partial \theta} + \frac{1}{2} \cdot i^2 \cdot \frac{\partial L_C}{\partial \theta} \tag{6.6}$$

The expression (6.6) defines the electromagnetic torque as the individual contribution of each phase of the SRG. Using a sinusoidal inductance model, the rate of change of inductance can be determined as shown in (2.31), (2.32), and (2.33).

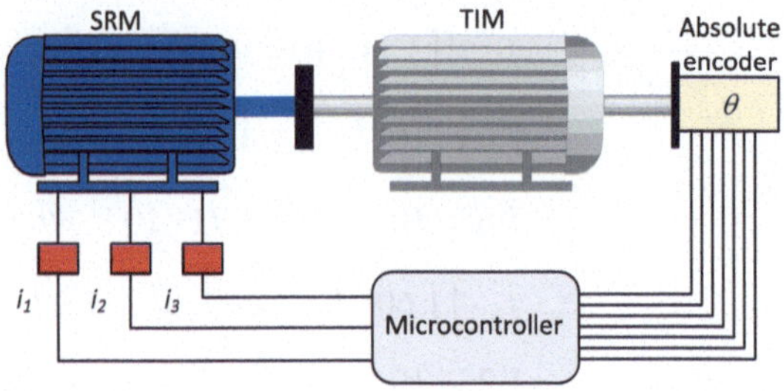

Fig. 6.8 System diagram for indirect torque detection

Consequently, the electromagnetic torque T_{EM} can be rewritten as:

$$T_{EM} = -2 \cdot B \cdot i_A^2 \cdot sen(4 \cdot \theta) - 2 \cdot B \cdot i_B^2 \cdot sen[4 \cdot (\theta - 60°)] - 2 \cdot B \cdot i_C^2 \cdot sen[4 \cdot (\theta + 60°)] \tag{6.7}$$

Thus, the procedure includes the design and development of hardware capable of calculating the instantaneous value of the electromagnetic torque based on the measurements of instantaneous currents in the windings i_A, i_B, and i_C, as well as the instantaneous angular position θ. These values enable the determination of the mechanical power required in the SRG shaft for energy generation, facilitating the calculation of the efficiency of the generator. Figure 6.8 illustrates the block diagram of the indirect torque measurement system [3].

The microcontroller (Fig. 6.8) measures the currents of the three SRG windings, which are acquired through three Hall-effect current sensors, indicated in red for each SRG phase. The signals from the sensors are appropriately conditioned and subsequently digitized by the microcontroller's analog-to-digital converter. This converter calculates the electromagnetic torque based on the instantaneous angular position θ of the rotor. Equation (6.7) is employed to compute the electromagnetic torque utilizing the sinusoidal inductance model [3].

However, the constant B in (2.35) is defined considering experimental inductance values. Specifically, the L_{MAX} values are expected to be determined by mapping the simulation surface, ensuring that the torque does not rely solely on the linear model. The microcontroller in Fig. 6.8 generates a PWM output signal, which, after filtering, can be read by the system's main controller, shown in Fig. 5.7. The instantaneous electromagnetic torque enables the calculation of the total torque on the SRG shaft, which is applied by the induction motor and can be determined using the Eq. (2.18). This torque value allows the calculation of the mechanical power required for the corresponding electrical power generation by the SRG, given by:

$$P_{mec} = T_M \cdot \omega \tag{6.8}$$

The input and output electrical powers of the generator can be determined, respectively, by:

$$P_1 = V_1 \cdot I_1 \tag{6.9}$$

and,

$$P_2 = V_2 \cdot I_2 \tag{6.10}$$

enabling the determination of the generator's efficiency through:

$$\eta_T = \frac{P_2}{P_1 + P_{mec}} \tag{6.11}$$

Thus, the procedures carried out allow the evaluation of other SRG parameters, facilitating the optimization of the generator's operating conditions to maximize energy production and enhance efficiency η_T.

6.4 Torque Detection Circuit

As the previous section validates the computational model using parametric regression to represent the inductances of the SRG, considering specific electrical quantities, this section presents the electromagnetic torque values provided by the proposed system shown in Fig. 6.8. The same tests performed for the excitation current and generated voltage are repeated to compare the electromagnetic torque obtained experimentally, through the proposed system, with the values provided by the computational model, which demonstrates relative accuracy. Figure 6.9 presents the simulated and experimental values of the electromagnetic torque of the SRG when subjected to excitation voltages of 100 V, 150 V, and 200 V [3].

Figure 6.9 shows that the values provided by the validated simulation and those obtained from the proposed system to detect electromagnetic torque, based on the angular position and instantaneous currents, exhibit relative proximity. The values displayed in Fig. 6.9 indicate deviations of 4.5%, 3.5%, and 7.5%, resulting in an average deviation of approximately 5%. Thus, the indirect torque detection system can be effectively used to estimate the total efficiency of the SRG, considering

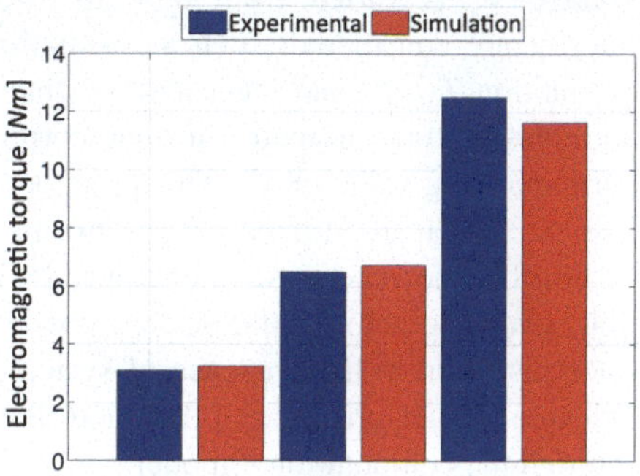

Fig. 6.9 Electromagnetic torque T_{EM} compared between the computational and experimental models

that the primary objective of this study is to enhance efficiency through power control [3].

6.5 Optimization of Phase Turn-Off Angle

Considering the computational model with the inductance surface represented by a continuous expression, the simulation is used to validate the torque estimator and to monitor the behavior of key SRG parameters under varying operating conditions. This procedure involves varying the turn-off angle of the SRG windings (θ_{OFF}) and the excitation voltage (V_1), allowing the evaluation of parameters such as excitation power (P_1), mechanical power (P_{mec}), total generator efficiency (η_T), electrical efficiency (η_E), and mechanical efficiency (η_M). Thus, the computational simulation generates the corresponding surfaces for each parameter [1].

For SRG power control aimed at improving total efficiency (η_T), some of these parameters are maximized while others are minimized. Power control employs the excitation voltage (V_1) as a variable manipulated. The surfaces generated by the simulation are analyzed to establish the relationship between V_1 and θ_{OFF}. Optimization techniques are applied to maximize or minimize an objective function. In this context, the parameters are adjusted to achieve optimal SRG efficiency at any excitation voltage provided by the controller [2].

Thus, it is necessary to apply an optimization algorithm to determine the relationship between the excitation voltage V_1 and the angle of extinction of the SRG phase windings θ_{OFF}. This relationship identifies the behavior of the system based on these two variables, allowing control of the generated power P_2 while maximizing efficiency. The optimization process to obtain the optimal values θ_{OFF} is performed after generating the surfaces of the relevant quantities. These surfaces are derived through simulation, exploring different values of θ_{OFF} and V_1. In addition to this optimization, the controller parameters shown in Fig. 6.10 are also adjusted to enhance performance [3].

In Fig. 6.10, two loops are highlighted: the power control loop and the optimization loop, which utilizes the function $\theta_{OT}(V_1)$. Specifically, the excitation voltage V_1 is varied from 0 V to 150 V in 10 V increments. Variables of interest are considered, such as excitation current I_1, torque required for power production T_{MEC}, and electrical, mechanical and total efficiencies. Each of these variables is stored in matrix format, correlating the optimal value of θ_{OFF} with the corresponding value of V_1. This process results in five surfaces for the variables of interest, which are utilized for optimization associated with power control.

For SRG power control, a PI controller is used, as shown in Fig. 6.10. The control objective is to achieve the desired power value P_{REF} within the minimized time intervals, even in the presence of system disturbances such as load variations. To enhance the performance of the controller, the parameters k_p and k_i are optimized through the computational model.

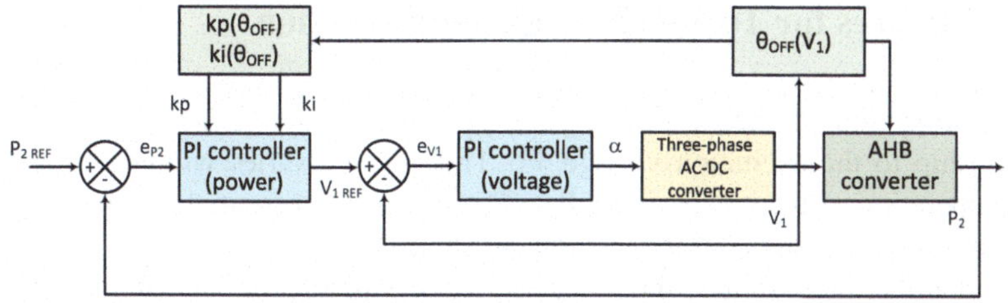

Fig. 6.10 Power control system diagram with optimized θ_{OFF} for efficiency maximization

In addition, the surfaces corresponding to the optimal θ_{OFF} values are used in formulating the objective function for the optimization system. The goal is to implement a function of the type:

$$f(x) = p_1 \cdot \left(\frac{1}{n_E}\right) + p_2 \cdot \left(\frac{1}{n_M}\right) + p_3 \cdot \left(\frac{1}{n_T}\right) + p_4 \cdot I_1 + p_5 \cdot T_{MEC} \quad (6.12)$$

Although weight tuning is challenging during the optimization process, the values of p_1 to p_5 must be set to maximize the overall efficiency of the switched reluctance generator. In this context, the function $\theta_{OT}(V_1)$ applies the most suitable value θ_{OFF} to the generator power converter, optimizing the efficiency based on the excitation voltage applied to the generator by the PI controller [3].

6.6 Sensitivity Analysis in the Operation and Control of the SRG

As sensitivity analysis techniques allow for the evaluation of how system input parameters influence output variables, they are applicable to the operation and control of the switched reluctance generator (SRG) [5]. In a closed-loop control system for speed regulation when operating as a motor, better performance is observed when the excitation voltage is used as the manipulated variable. In this study, the same analysis is conducted for generator operation to determine whether output power control is more effective when acting on the excitation voltage or the phase turn-off angle.

To this end, sensitivity analysis is applied to evaluate the influence of V_1 and θ_{OFF} on the generated power. Similarly, the power controller is evaluated by identifying the impact of the gains of the PI controller (k_p and k_i) on the integral of absolute error (IAE). The sensitivity analysis applied to the power controller parameters significantly aids the controller optimization algorithm by facilitating the adjustment of input variable step sizes during iterations [2].

6.7 Surfaces for Turn-Off Angle Optimization

The optimization of the turnoff angle of the windings (θ_{OFF}) is based on its relationship to the excitation voltage V_1. This voltage varies within a range due to the operation of the power controller. Therefore, it is essential to determine the optimal turn-off angle θ_{OT} for each voltage value [3].

To identify these values, the surfaces of the variables influencing the SRG's efficiency are mapped. The total efficiency reaches approximately 80% for an excitation of 150 V and a speed of 1000 rpm. However, factors such as reducing the input mechanical power, decreasing the excitation power, and increasing the generated power contribute positively to improving the SRG's total efficiency.

Thus, it is crucial to understand the behavior of these variables in response to variations in θ_{OFF} and the excitation voltage V_1. Consequently, the surfaces of these variables are generated as functions of the turn-off angle and the excitation voltage. Figure 6.11 illustrates the surfaces obtained using the computational model for the excitation power P_1 and the generated power P_2, considering variations in the turn-off angle from 2° to 40° and in the excitation voltage V_1 from 150 V to 264 V.

Figure 6.12 presents the surfaces of the mechanical input power and the mechanical efficiency of the SRG. In all simulations conducted to generate these surfaces, the phase turn-on angle θ_{ON} is fixed at 3°. This delay ensures that, when the coil is turned off at $\theta_{OFF} = \theta_{ON} + 30°$, the current begins to decrease at a point where the inductance is higher, maximizing the generation stage. Consequently, the angle θ_{OFF} is varied starting from 2°, ensuring a minimum conduction window of 5° [2].

Finally, Fig. 6.13 presents the surfaces of the electrical efficiency and the total efficiency of the generator. The electrical and mechanical efficiencies represent the relationship between the generated power and the excitation and mechanical input powers, respectively. These relationships are described by Eqs. (6.13) and (6.14).

$$n_M = \frac{P_2}{P_{MEC} + P_2} \tag{6.13}$$

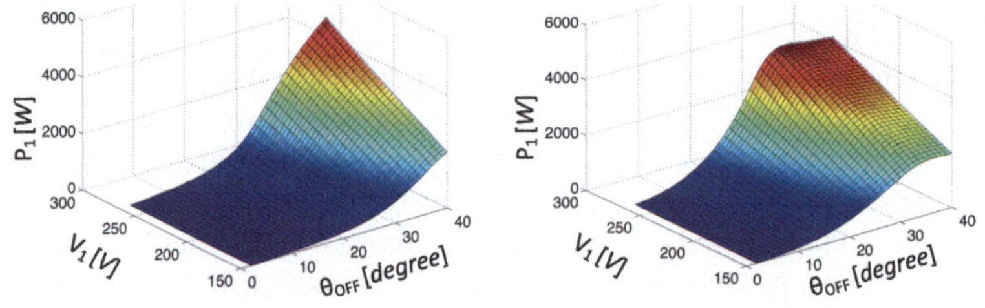

Fig. 6.11 Surfaces of excitation power and generated power as functions of θ_{OFF} and V_1 variations

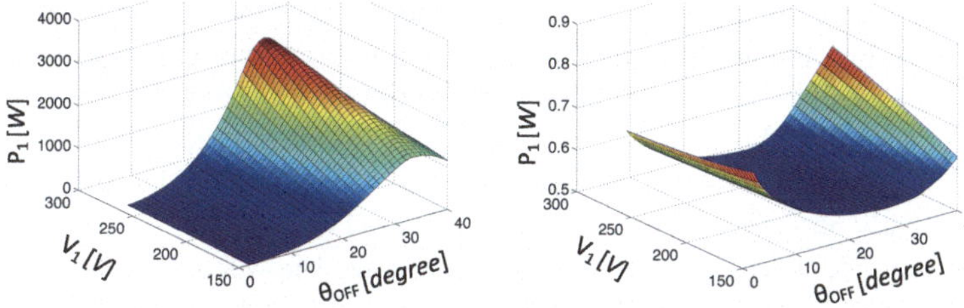

Fig. 6.12 Surfaces of mechanical input power and mechanical efficiency as a function of θ_{OFF} and V_1 variations

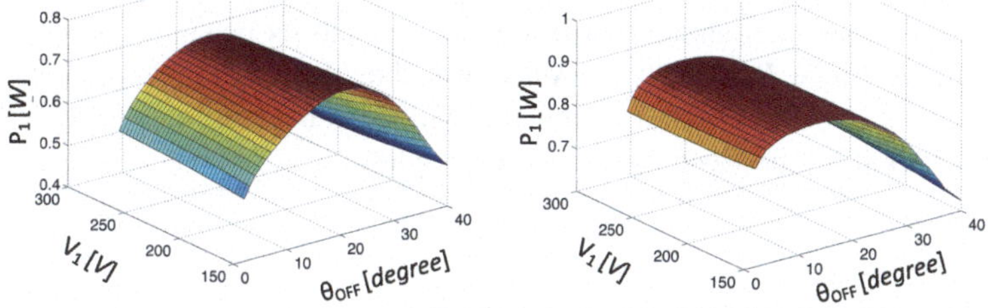

Fig. 6.13 Surfaces of electrical efficiency and total efficiency of the SRG as a function of θ_{OFF} and V_1 variations

$$n_E = \frac{P_2}{P_1 + P_2} \tag{6.14}$$

Parameters n_M and n_E are considered to contribute to the identification of the relationship $\theta_{OT}(V_1)$, since the objective is to minimize the input power while maximizing the output power [3].

6.8 Construction of the Proposed Turbine Model

The modeling of the turbine is a crucial step in understanding and improving the operation of the wind generation system. The adopted methodology involves defining the input parameters and calculating variables such as the power coefficient C_P, the mechanical power of the wind turbine P_t, and the developed torque T_t. The first step is to define the input parameters as they directly influence the turbine's performance. The constants of the model include the diameter of the swept area of the rotor A, the pitch angle β, and the air density ρ, while the variables involved are the wind speed v_v and the turbine rotational speed ω_t. An accurate definition of these parameters is essential to ensure the precision of the wind generation model [1].

After defining the input parameters, the output variables are calculated. Initially, the power coefficient C_P is determined using Eq. (3.10). This calculation is fundamental to assess the efficiency and performance of a wind turbine. The value of C_P varies depending on the wind speed and the operational conditions of the turbine, including the angle of inclination of the blade β and the rotational speed ω_g. The power coefficient has a theoretical maximum limit of 0.59, known as the Betz limit. However, this limit is idealized and real-world turbines exhibit lower C_P values. Therefore, a maximum power coefficient value $C_{P_{\text{Max}}}$ is defined, reflecting values typically observed in practical wind turbines [1].

Using the calculated value of C_P, the mechanical power of the turbine P_t is determined using Eq. (3.8), and its torque T_t is calculated with Eq. (3.12). This requires the wind speed v_v, which can be treated as either a constant or a vector of velocity values. To accurately represent the real performance of the SRG in wind energy generation, real-world data v_v are collected. After constructing the wind turbine model, simulations are performed to evaluate its behavior under varying v_v conditions. Since P_t depends not only on v_v but also on ω_t, a maximum power point tracking (MPPT) method is implemented. This method aims to identify the optimal relationship between v_v and ω_g, known as λ^*.

6.9 Maximum Power Point Tracking

In addition to variations in the designs of wind turbine shafts, which can be horizontal or vertical, there are also turbine systems with fixed or variable speeds. This book adopts the horizontal-axis, variable-speed turbine model, which allows for better energy utilization. Determining the maximum power point of a wind turbine involves optimizing ω_t with respect to v_v to achieve the maximum mechanical power. Using the definition of λ^*, Eq. (3.9) can be rewritten in terms of ω_t, enabling the calculation of ω_t for each v_v value, as expressed by:

$$\omega_t = \frac{\lambda^* \cdot v_v}{R_t} \tag{6.15}$$

Since the turbine radius R_t is constant, it is possible to determine ω_t knowing the value of v_v. As the wind speed varies over time, once λ^* is known, ω_t can be calculated to achieve $C_{P_{\text{Max}}}$, thus extracting the maximum possible mechanical power. The coupling between the wind turbine and the SRG is modeled to establish a direct relationship between the speed and torque of both the turbine and the generator. Since wind turbines typically operate at low speeds, gearboxes are commonly used to increase speed to values within the operating range of the generator. However, the SRG has the ability to operate at low rotational speeds, unlike other generator technologies frequently employed in wind energy generation, such as induction generators [1].

Due to the SRG's ability to operate at low speeds, the gear ratio G can be lower than that used in other machines. In some cases, this ratio can be eliminated, allowing the wind turbine and the generator to operate at the same speed. Gearbox

Table 6.4 Characteristics of the simulated wind turbine

Parameter	Value
R_t	1.6 m
ρ	1.22 kg/m^3
β	0°
$C_{P_{\text{Max}}}$	0.45

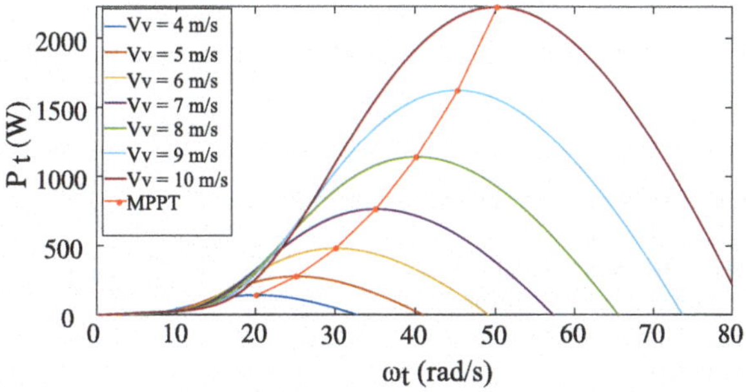

Fig. 6.14 Turbine power as a function of wind speed and turbine rotation with maximum power extraction

modeling requires knowledge of parameters such as inertia J_t, stiffness K_s, and shaft damping constant D_s. However, due to the difficulty of accurately representing these parameters and the ability of the SRG to operate without a gearbox in certain scenarios, the ideal value of G for the proposed model is evaluated based on the nominal speed of the SRG.

The mathematical model of the wind turbine is formulated according to Eqs. (3.7) to (3.12), which serve as the basis for developing the computational model. Using this computational model, simulations are conducted to analyze the behavior of the wind turbine. For a turbine radius of $R_t = 1.6$ m, the mechanical power generated reaches 2000 W at a wind speed of 10 m/s, ensuring that the power is appropriately scaled for the generator used in the simulation. The angle of rotation of the blade is set to $\beta = 0°$ to achieve the maximum power coefficient, $C_{P_{\text{Max}}} = 0.45$. After defining the input parameters for the wind turbine model, simulations are performed with the wind speed v_v varying between 4 m/s and 10 m/s [1].

Table 6.4 presents the characteristics of the wind turbine, while Fig. 6.14 illustrates the behavior of the turbine power P_t for different wind speed values. It is observed that as v_v increases, the power P_t also increases. However, the power extracted by the turbine does not depend exclusively on v_v, since specific values of ω_t allow maximum power extraction from the wind turbine for each value of v_v, as shown in Fig. 6.14 by the ascending red curve [1].

For the implementation of the maximum power point tracking (MPPT) algorithm, the pitch angle of the blade β is initially evaluated. This angle is defined as the angle between the turbine blades and the line perpendicular to its axis. When $\beta = 0°$, the blades are aligned with the direction of the wind, and increasing this value causes the blades to tilt relative to the direction of the wind. Therefore, β is set to 0°, allowing the achievement of $C_{P_{Max}}$. With β constant, C_P varies as a function

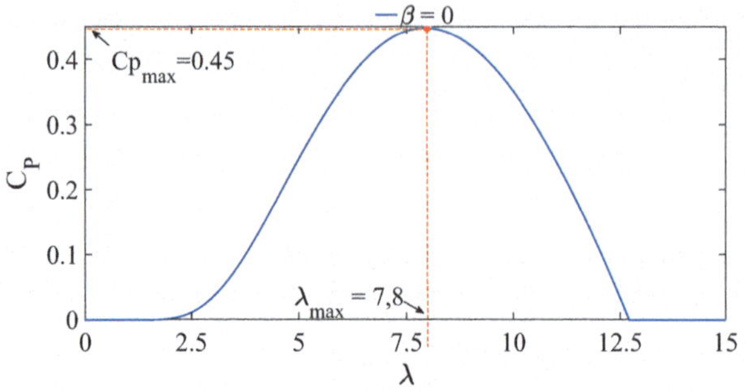

Fig. 6.15 Power coefficient as a function of λ

Fig. 6.16 Mechanical power and turbine rotational speed under abrupt wind speed variations

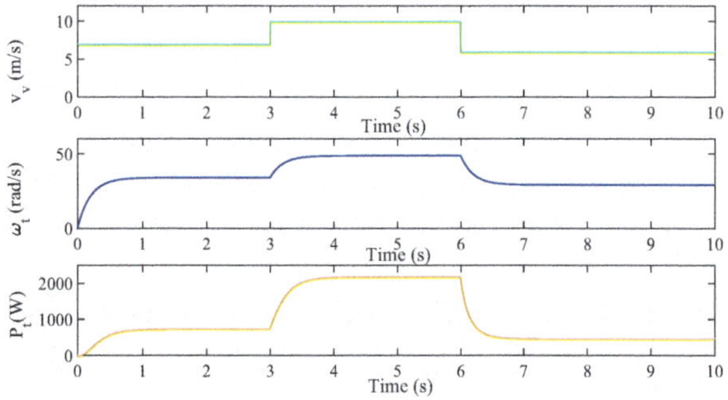

of λ. Thus, the value of λ that results in $C_{P_{Max}}$ is determined by analyzing the curve of C_P as a function of λ, as shown in Fig. 6.15 [1].

Thus, the values $\lambda^* = 7.8$ and $C_{P_{Max}} = 0.45$ are defined. Using these values, ω_t is calculated for each value of v_v using Eq. (6.15). The wind turbine is then simulated with the implemented MPPT algorithm. The simulation is carried out with $v_v = [7, \ 10, \ 6]$ m/s, with intervals of $3s$, to evaluate the behavior of the model under abrupt changes in v_v. This scenario reflects real-world conditions in which wind turbines operate with sudden variations in wind speed.

Figure 6.16 illustrates the simulated behavior for abrupt variations in v_v. The total simulation time is $t = 10$ s. At $t = 3$ s, v_v increases from 7 m/s to 10 m/s, and both P_t and ω_t are observed to increase almost instantaneously. Similarly, at $t = 6$ s, v_v decreases from 10 m/s to 6 m/s, and both P_t and ω_t decrease proportionally. Therefore, the model behaves as expected in response to changes in v_v.

6.10 Simulation of Turbine-Generator Coupling

The speed ranges at which wind turbines are typically operating are considered low compared to the operating speed ranges of conventional generators. Given the SRG's ability to operate at low rotational speeds, a speed multiplication ratio of

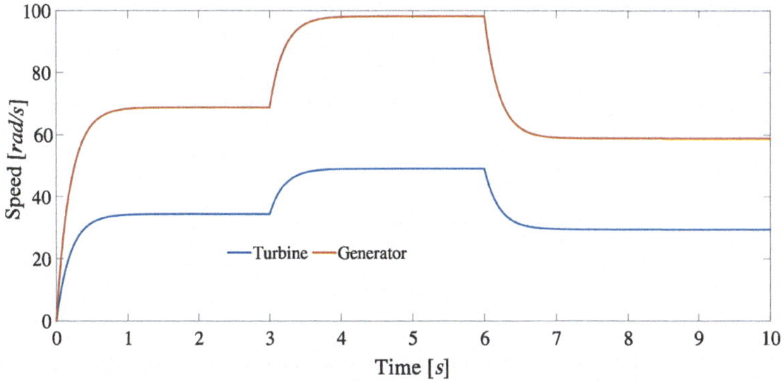

Fig. 6.17 Turbine and generator rotational speed with multiplication factor $G = 2$

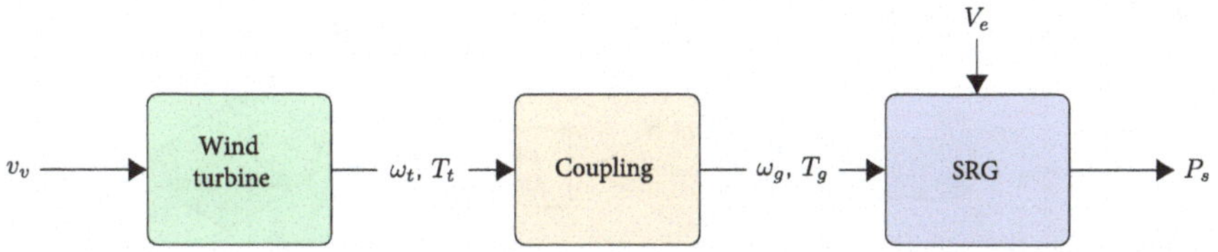

Fig. 6.18 Flowchart of the wind generation system model

$G = 2$ was defined, which was deemed ideal for the proposed model. Since one of the objectives of this book is to evaluate the switched resistance generator (SRG) in low-speed operation, $G = 2$ was the minimum feasible value capable of generating energy. Figure 6.17 presents the input and output speeds of the coupling, represented by ω_t and ω_g, respectively, for $v_v = [7,\ 10,\ 6]$ m/s with intervals of 3 s [1].

After constructing the models for the wind turbine, the coupling system, the SRG, and collecting real wind speed data (v_v), it becomes feasible to integrate each block to form the model of the complete wind generation system. Figure 6.18 presents the flowchart implemented for constructing this model. For each value of v_v, the wind turbine block calculates the rotational speed of the turbine ω_t and the developed torque T_t. The coupling block connects the low-speed side of the wind turbine to the high-speed side of the generator. The SRG block receives the rotational speed ω_g, the torque T_g from the coupling system, the excitation voltage V_e supplied by the AHB converter and outputs the electrical power P_s.

With the wind turbine model established, simulations are carried out to observe the behavior of the SRG under various v_v inputs, enabling the evaluation of its performance with real v_v data. Finally, optimization techniques are applied to enhance efficiency by maximizing the generated power. The primary variables manipulated to achieve maximum generated power and system efficiency are the excitation voltage V_e, the turn-on angle of the coils θ_{on}, and the turn-off angle of the coils θ_{off}.

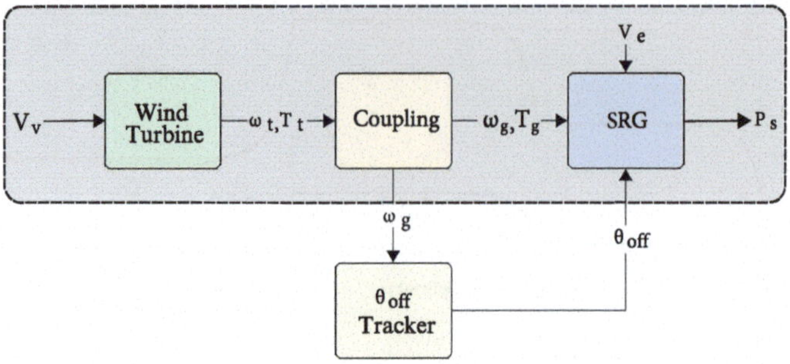

Fig. 6.19 Flowchart of the wind power generation system model and optimization process with θ_{off} tracking

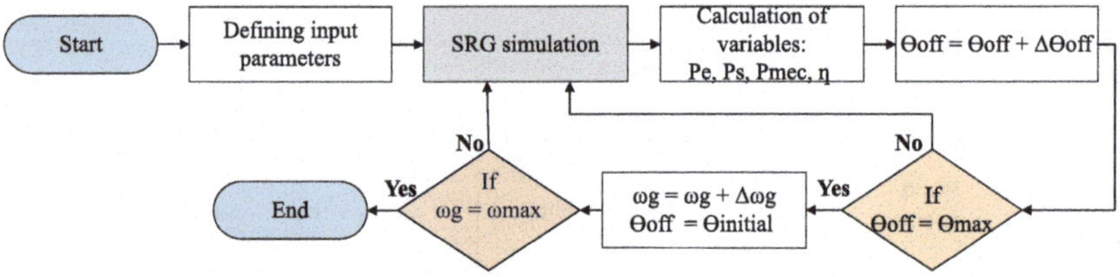

Fig. 6.20 Flowchart of the θ_{off} tracking action

The methodology adopted in this study involves keeping V_e and θ_{on} constant while tracking the θ_{off} values that maximize the power generated throughout the entire SRG operating speed range. After defining the excitation voltage V_e, simulations are performed to identify the optimal value of θ_{on}, considering both the generated electrical power and efficiency. Once this value is determined, an additional block, called the θ_{off} tracker, is incorporated into the model, as illustrated in Fig. 6.19.

The θ_{off} tracker block receives the ω_g value from the coupling block and determines the optimal θ_{off} based on this input. To achieve this, the curves of input electrical power P_e, output electrical power P_s, and mechanical power P_{mec}, along with the efficiency curve η, must be analyzed considering the values of θ_{off} and ω_g. The proposed methodology for tracking θ_{off} is illustrated in Fig. 6.20 [1].

Initially, the initial values for the excitation voltage V_e, the angles θ_{on} and θ_{off}, and the generator rotational speed ω_g are defined. With these parameters configured, the first simulation is performed, allowing the calculation of the mechanical power P_{mec}, the electrical input power P_e, and the electrical output power P_s, which can be obtained, respectively, through the following expressions:

$$P_{mec} = T_M \cdot \omega_g \tag{6.16}$$

$$P_e = V_e \cdot I_e \tag{6.17}$$

$$P_s = V_s \cdot I_s \tag{6.18}$$

With the power values, the efficiency can be determined using the following expression:

$$\eta = \frac{P_s}{P_e + P_{mec}} \tag{6.19}$$

After calculations, the value of θ_{off} is incremented by $\Delta\theta_{off}$ until it reaches the maximum value θ_{max}. In this way, the values of P_{mec}, P_e, P_s, and η are obtained for each θ_{off}. However, these values also need to be calculated for each ω_g. To achieve this, ω_g is incremented by $\Delta\omega_g$ until it reaches the maximum value ω_{max} [1].

At the end of this process, the power and efficiency surfaces are generated in the entire range of θ_{off} and ω_g analyzed. These surfaces make it possible to identify the values θ_{off} that maximize power and efficiency for each speed analyzed. However, the goal is to derive a polynomial function that can perform this calculation at any speed. Once this function is determined, it is incorporated into the simulation through the θ_{off} tracker block shown in Fig. 6.20, with the purpose of generating the θ_{off} reference for the AHB converter.

6.11 Optimization Process Applied

If the goal is to implement control of the SRG output voltage V_s in conjunction with the θ_{off} tracker, a PID controller is incorporated into the simulation. The variables manipulated to control V_s are: (i) the excitation voltage V_e, (ii) the turning angle of the struts θ_{on}, and (iii) the turning angle of the windings θ_{off}. Initially, the optimal value θ_{on} is sought to positively influence the generated power P_s, the efficiency η, and the torque ripple T_R. Under these conditions, the conduction window $\theta_D = 30°$ is maintained, and θ_{on} is reduced incrementally until the established conditions are met. For this analysis, θ_{on} remains fixed while a sensitivity analysis is performed considering the variables: (i) excitation voltage V_e, (ii) turn-off angle θ_{off}, and (iii) generator rotational speed ω_g [3].

The optimization process identifies $\theta_{on} = -10°$ as the optimal value. This value is fixed and the influence of the variables manipulated on P_s, η, and T_R is evaluated. The sensitivity indices indicate that V_e has the greatest influence on P_s and η, while θ_{off} has the most significant impact on T_R. Consequently, V_e is selected as the variable manipulated to control V_s, ensuring maximum system efficiency.

The PID controller is chosen for its straightforward implementation and low computational cost. It is used to control V_s, with V_e as the manipulated variable. Although various studies address tuning processes for this type of controller, optimization methods enable superior performance [1]. In this work, the Nelder–Mead simplex algorithm is employed, a deterministic optimization method. For implementing the PID controller in the simulation, the reference voltage $V_{s_{\text{ref}}}$ is defined, and the gains k_p, k_i, and k_d, corresponding to the proportional, integral and derivative gains of the controller, respectively, are tuned. The evaluation function is defined as:

$$f_{V_s}(k_p, k_i, k_d) = \frac{1}{3} \cdot \left[\int_0^t \left| \frac{e_{V_s}(t)}{V_{s_{\text{ref}}}(t)} \right| dt + \left| \frac{V_s(t_M) - V_s(\infty) + e_{V_s}(\infty)}{V_{s_{\text{ref}}}(\infty)} \right| \right] \tag{6.20}$$

In (6.20), the normalized minimization is defined as comprising the error area, represented by the integral of absolute error (IAE), the overshoot, given by $V_s(t_M) - V_s(\infty)$, and the steady-state error, expressed as $e_{V_s}(\infty)$. After determining the optimized controller gains, simulations are performed to apply perturbations to the system and evaluate its performance in controlling V_s. This step is crucial to validate the proposed wind generation model and serves as a foundation for decisions on the feasibility of SRG implementation [2].

Simulations are carried out for different values of v_v. Initially, v_v is considered constant and, subsequently, variations are introduced to analyze the behavior of the system under abrupt changes over short time intervals. Since wind is a dynamic variable subject to constant change, and given that these variations directly impact the generated power, it is essential that the generator demonstrate operational capability over a wide range of v_v values. Real data analysis v_v is necessary to evaluate the actual performance of the SRG in wind generation applications. These data are obtained from meteorological databases that cover stations in Brazil or from atlases that analyze the wind potential in specific regions of the country. Based on this information, suitable locations for wind turbine installation are selected.

The data in v_v may require adjustments, as measurement heights vary between meteorological stations. After adapting the data to the specific heights considered in the analysis, the regions to be studied are selected. To evaluate SRG performance under low, medium, and high conditions v_v, data from three different regions of Brazil are used, encompassing a wide range of v_v profiles. The validation of the model used in this study involves comparing simulated data with experimental results and simulations from other studies using the same switched resistance machine. In this context, the validated models include both the wind turbine and the SRG.

The model used to represent the horizontal-axis wind turbine is widely recognized and is frequently used in various studies. This is because the expression for calculating the power of the wind turbine, presented in (3.8), is directly derived from physical principles, establishing the relationship between the conversion of wind kinetic energy into mechanical energy and, subsequently, into electrical energy. The precision of the wind turbine model depends on the proper definition of its parameters and the calculation of C_P. The methods for calculating C_P are based on the parameters β and λ, with several approaches available to represent the coefficients using the expressions (3.10) and (3.11).

These methods are validated through simulations and experimental tests. In addition, other studies corroborate the effectiveness of this calculation method in

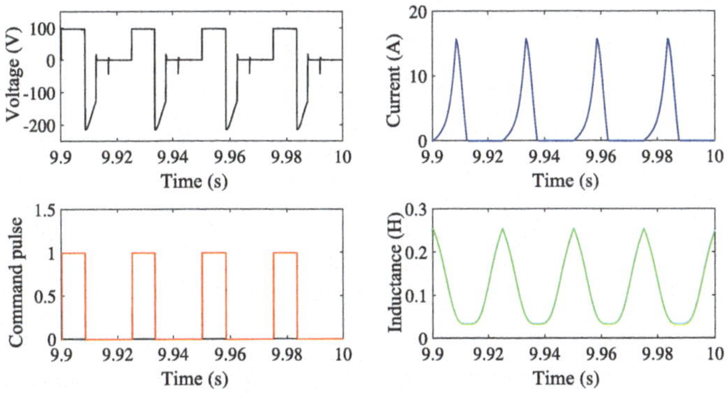

Fig. 6.21 Behavior of voltage, current, voltage pulse, and inductance of Phase **A**

determining the coefficient C_P. The simulations conducted for the wind turbine in this study indicate that increasing the parameter β reduces both the total power P_t and the coefficient C_P. They also reveal the existence of a specific λ value that maximizes the extracted power. The SRG computational model was developed based on expressions (2.1) to (2.30). The SRG parameters, presented in Table 6.1, reflect the design and construction of an SRG with a configuration 6×4 [1]. The accuracy of the SRG model is directly linked to the representation of its inductances. In this study, the inductance representation model described in Chap. 4 is also used, using the parametric regression process.

The simulations were carried out with an excitation voltage of $V_e = 100$ V, maintaining a constant rotational speed of 600 rpm, and the windings' switching angles were set at $\theta_{on} = 0°$ and $\theta_{off} = 30°$. In addition, a resistive load of $R_{carga} = 33\ \Omega$ was connected to the generator output. Figure 6.21 illustrates the behavior of the voltage, current, command pulse, and inductance of phase **A**. The IGBT command pulse is applied during the period when the inductance is decreasing, enabling the machine to operate as a generator. Figure 6.22 shows the command pulse of Phase **A** and its inductance, with the inductance curve magnified by a factor of seven for better visualization.

In Fig. 6.22, it is observed that the pulse is applied when the inductance reaches its maximum value, at position $\theta_{on} = 0°$, and is turned off when the inductance reaches its minimum value, at position $\theta_{off} = 30°$, maintaining a conduction window of $\theta_D = 30°$. This configuration is widely recognized in the literature as optimal for switched reluctance machines (SRM) with a configuration 6×4. To operate as a generator, the control pulse for the IGBT switches must be applied during the period when the inductance is decreasing, as illustrated in Fig. 6.22. The SRM was simulated with an excitation voltage of $V_e = 100$ V and a rotational speed of $\omega_g = 600$ rpm. Figure 6.23 presents the behavior of the voltage, current, control pulse, and inductance.

In general, most studies initiate machine excitation with $\theta_{on} = 0°$ and $\theta_{off} = 30°$, marking the start of the generation phase, during which the machine achieves

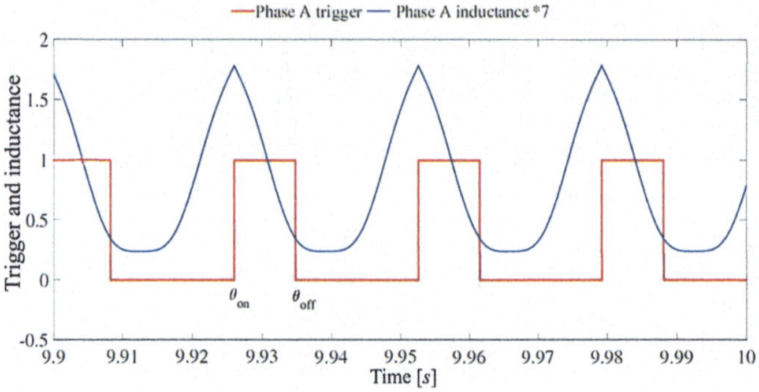

Fig. 6.22 Trigger and inductance of Phase **A**

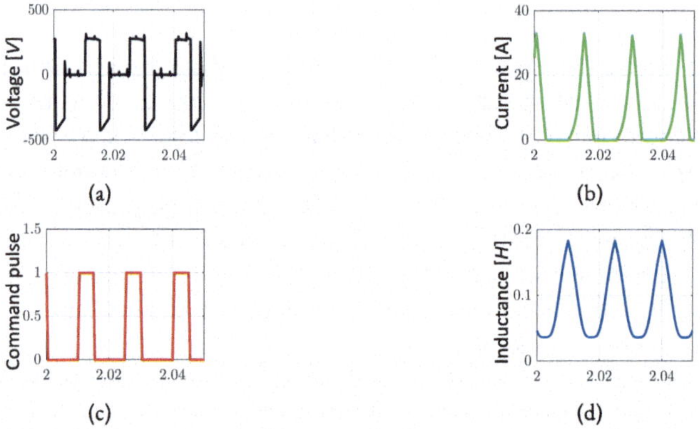

Fig. 6.23 Behavior of one phase of the machine: (**a**) voltage, (**b**) current, (**c**) trigger, and (**d**) inductance

Fig. 6.24 Mechanical power, electrical input power, and electrical output power of the SRG

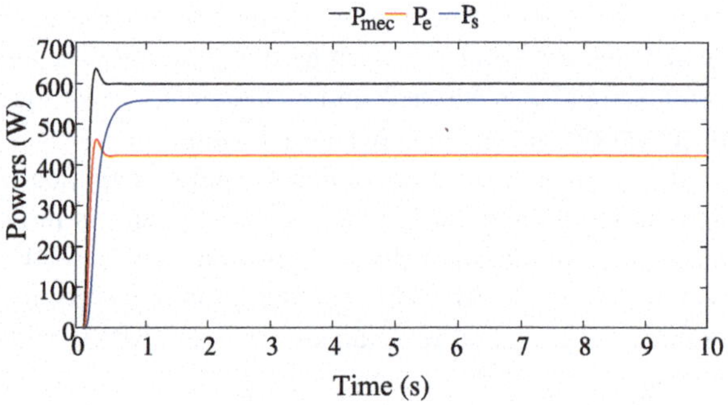

an output voltage higher than the input voltage. In this work, the generator achieved 216 V during the generation phase with $V_e = 100$ V, meeting the expected standards. Furthermore, the output power P_s exceeded the input power P_e in all simulations carried out, as shown in Fig. 6.24.

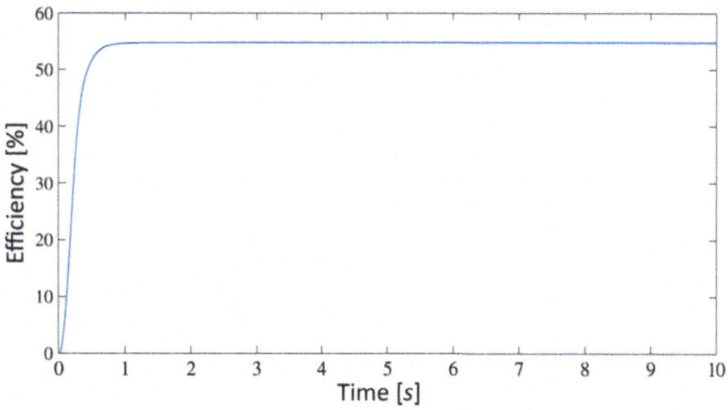

Fig. 6.25 Efficiency of the SRG

Table 6.5 Characteristics of the simulated wind generation system

Parameter	Value
G	2
V_e	180 V
θ_{on}	0°
θ_{off}	30°

Figure 6.24 illustrates the behavior of mechanical power (P_{mec}), electrical input power (P_e), and electrical output power (P_s). With an applied power of $P_e \approx 400$ W, the machine achieves $P_{mec} \approx 600$ W and $P_s \approx 559$ W, both exceeding the power consumed by the excitation source. Figure 6.25 shows the efficiency curve of the simulated SRM, calculated using Eq. (6.19) [1].

Following the construction and validation of the wind turbine, coupling, and SRG models, it is possible to interconnect each block to form the complete model of the wind generation system, as depicted in Fig. 6.18. With the complete model, the SRG's performance under constant wind speeds can be evaluated. Table 6.5 provides some of the parameters used for this simulation, while the remaining parameters for the wind turbine and the SRG are consistent with those listed in Tables 6.1 and 6.4.

The performance of the SRG is evaluated while maintaining v_v constant at 7 m/s. Figure 6.26 illustrates the variables v_v, ω_t, and ω_g, showing that when v_v remains constant, both ω_t and ω_g remain stable. A similar pattern is observed in Fig. 6.27, which shows the powers P_e, P_{mec}, P_s, and the efficiency η. In subsequent simulations, the excitation source is connected to the generator at $t = 1$ s, allowing the machine to reach steady state operation. The objective of the simulations with constant v_v is to analyze the behavior of the wind generation system under stable conditions.

To perform simulations with real v_v data, it is necessary to obtain measurements from meteorological stations located at potential sites for wind turbine installation. Reliable sources include specialized atlases that assess wind potential through the collection and analysis of v_v data. These datasets enable precise estimates of wind energy generation in different regions of the country. The use of real data v_v aims to evaluate the performance of the SRG in the generation of wind energy for specific regions, considering various scenarios v_v.

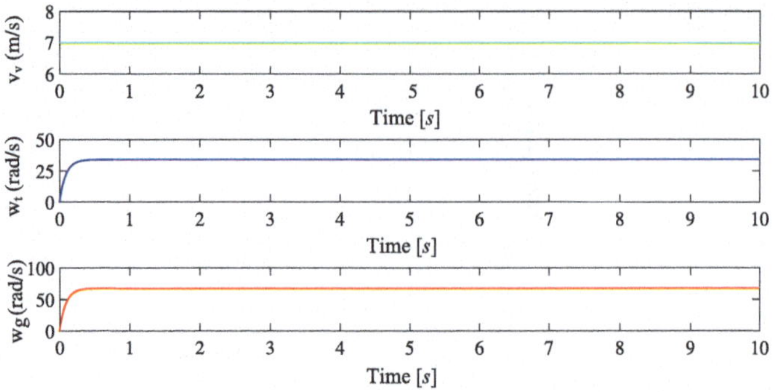

Fig. 6.26 Rotational speeds of the turbine and SRG for constant wind speeds

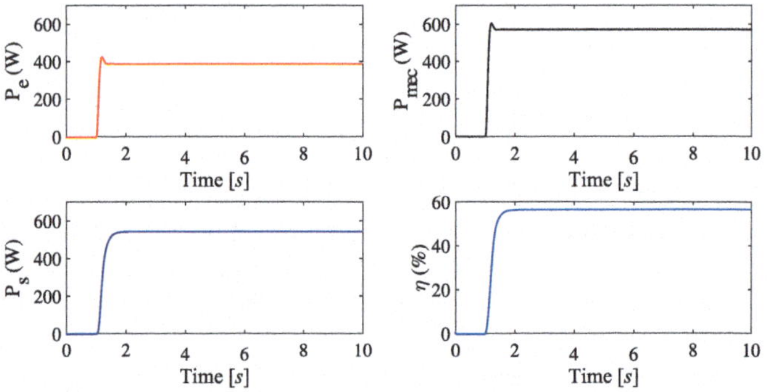

Fig. 6.27 Powers and efficiency of the SRG for constant wind speeds

From the data provided by the atlases, parameters such as annual average wind speeds $\overline{V_v}$ at heights of 30 m, 50 m, 80 m, 100 m, 120 m, 150 m, and 200 m can be extracted. In addition, metrics such as power density and diurnal profiles are available. The diurnal profile, also referred to as the **typical day**, reflects variations in v_v throughout the day, calculated as the arithmetic mean of wind speeds recorded at specific hours over the course of a year at a height of 100 m. The analysis of v_v behavior during the typical day can be performed using latitude and longitude data for the specific location.

Typically, these atlases are developed with a spatial resolution of 5 km × 5 km. In selecting locations for simulations, it is important to consider regions with distinct profiles v_v, including those classified as low, medium, and high wind speeds for energy generation. Table 6.6 presents the average annual wind speeds $\overline{V_v}$, measured at a height of 100 m, for the three selected locations [1].

From the selected locations, daily wind speed data series were extracted for each city. Subsequently, the monthly averages for each city were calculated and simulated over a 12-month period. Figure 6.28 illustrates the wind speed profiles for the three regions, representing the speed profiles analyzed. To perform the simulations, the obtained data were interpolated, enabling the creation of a dataset with multiple

Table 6.6 Locations for simulation with real data

#	Region	Latitude	Longitude	$\overline{v_v}$
Location 1	Deer Lake, Terra Nova/Canada	49.11°N	57.22°W	5.4 m/s
Location 2	Sorocaba, Sao Paulo/Brazil	−22.21°S	−49.94°W	7.1 m/s
Location 3	Faro, Algarve/Portugal	38.57°N	−7.90°W	8.4 m/s

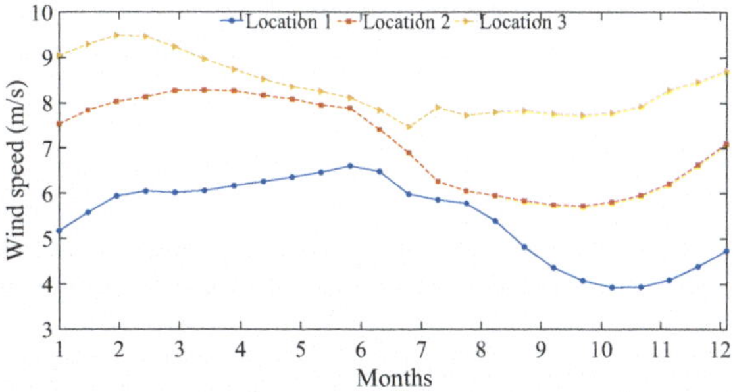

Fig. 6.28 Wind profile of the three chosen locations

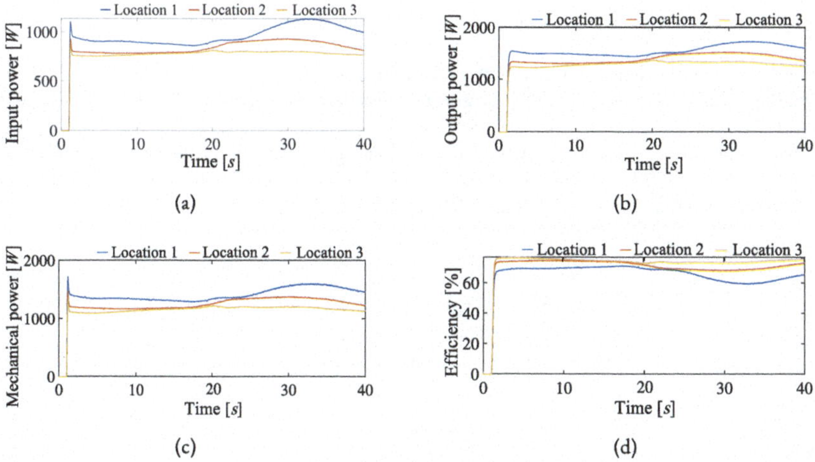

Fig. 6.29 Powers and efficiency of the SRG with independent excitation: (**a**) electrical input power, (**b**) electrical output power, (**c**) mechanical power, and (**d**) efficiency

values for each selected location. Figure 6.29 presents the powers P_e, P_s, P_{mec}, and the efficiency η for the three speed profiles analyzed. The SRG demonstrated the ability to generate energy throughout the analyzed speed range, achieving powers greater than 1000 W. Furthermore, the efficiency η remained above 60% throughout all the simulations performed.

In Fig. 6.29, the influence of v_v on the system's power outputs is observed. However, despite an increase in v_v, the power levels show a reduction. This behavior occurs due to the operation of the maximum power point tracking algorithm (MPPT), which adjusts the rotational speed ω_g to match the increase in v_v, resulting in a shorter excitation time for the generator. As the windings are excited for shorter periods, the electrical input power P_e decreases, resulting in a reduction in the mechanical power P_{mec}.

To optimize power generation, P_e must increase in proportion to the increase in rotational speed ω_g. To achieve maximum system efficiency, the θ_{off} tracking strategy is implemented. This approach involves maintaining fixed values for θ_{on} and V_e while determining the optimal value θ_{off} that maximizes power generation.

References

1. L.D.S. Morais et al., Enhancement of the performance of Switched reluctance generators in low wind speed conditions using advanced tracking techniques. Int. J. Energy Res. **2024**(1), 5541150 (2024)
2. M.R. Cunha Reis et al., Optimized techniques for driving and control of the switched reluctance motor to improve efficiency. Control Eng. Pract. **90**, 1–18 (2019)
3. W.R.H. Araujo et al., Efficiency enhancement of switched reluctance generator employing optimized control associated with tracking technique. Energies **14**(24), 8388 (2021). https://doi.org/10.3390/en14248388
4. F.A. Gomes et al., Heuristic regression method for descriptive data analysis, in *2016 IEEE 16th International Conference on Environmentand Electrical Engineering (EEEIC)* (IEEE, 2016), pp. 1–5
5. V.M. Gomes et al., Analytical method for calculating the sensitivity index of system parameters, in *2017 CHILEAN Conference on Electrical, Electronics Engineering, Information and Communication Technologies (CHILECON)* (IEEE, 2017), pp. 1–6

Chapter 7
Applications in Control and Energy Efficiency with Switched Reluctance Motors

7.1 Speed Control with Fixed Switching Angle

The switched reluctance motor is controlled by the classical proportional-integral-derivative (PID) controller. The PID controller is chosen for its simplicity in implementation and its analytical nature with low computational cost. The SRM computational model is developed as shown in Fig. 2.20, utilizing the inductance profile obtained through parametric regression, represented in Fig. 7.1 by the blocks **switchable reluctance motor**, **inductance**, and **switching** [1].

In the SRM simulation model, essential computational routines are implemented for the drive and control modeling, namely: (i) the AC-DC converter, representing the fully controlled three-phase rectifier (FCTR), (ii) the PID controller, and (iii) the hysteresis controller. Thus, speed control is achieved through the conduction angle of the electronic switches α of the AC-DC converter, represented in Fig. 7.1 by the highlighted blue line. The speed control signal α, illustrated in Fig. 7.1 by the red line, directly influences the excitation voltage V_{exc} of the SRM [2].

With fixed switching angles in the half-bridge converter, the current conduction window in the coil is maintained at 30°, as recommended in the literature for the SRM 6 × 4. Hysteresis control is employed to ensure that the starting current of the motor does not exceed its manufacturing limits. With reference to the peak current $I_{p_{ref}}$, the control signal of the hysteresis controller S_I is multiplied by the signal to turn off the upper switches $S_{\theta_{off}}$ of the half-bridge converter, calculated in the **switching** block, producing the signal applied to the gates $G_{Q_{2\varsigma-1}}$, as shown in Fig. 2.9 [1].

The PID controller is a linear model applied to a nonlinear system. Therefore, it is necessary to establish the control operating point for the reference speed $\omega_{ref}(t)$ (setpoint) and then adjust the controller gains k_p, k_i, and k_d. Several PID controller tuning methods are described in the literature. Among these methods, those using optimization processes are preferred due to their superior results in the following scenarios: (i) the motor operates without load, (ii) with mechanical load applied to

W. P. Calixto et al., *Control and Tracking Techniques for Switched Reluctance Machines*, Power Systems, https://doi.org/10.1007/978-3-031-86727-9_7

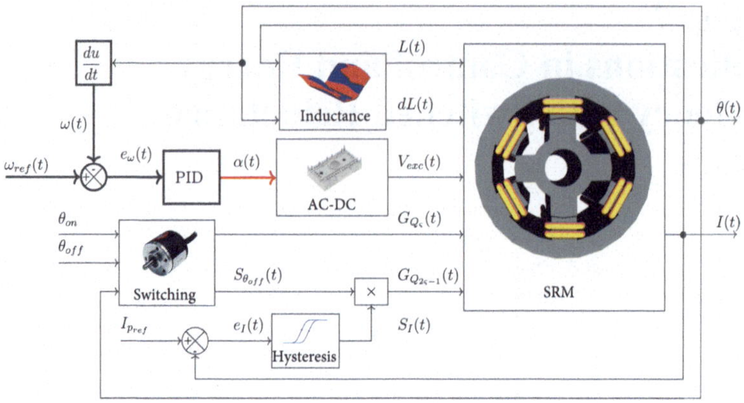

Fig. 7.1 Excitation voltage speed control system with fixed switching angles

the motor shaft, and (iii) with variation of the setpoint. For this, the hybrid GA-QNM optimization algorithm is used, using the evaluation function given by (7.1) to determine the optimal speed adjustment.

$$f_\omega(k_p, k_i, k_d) = \frac{1}{3} \cdot \left[\int_0^\infty \left| \frac{e_\omega(t)}{\omega_{ref}(t)} \right| dt + \left| \frac{\omega(t_M) - \omega(\infty) + e_\omega(\infty)}{\omega_{ref}(\infty)} \right| \right] \tag{7.1}$$

where t_M represents the instant when ω reaches its maximum value at t. The expression (7.1) describes the normalized minimization, considering the error area evaluated by the integral of the absolute error (IAE), the overshoot given by $\omega(t_M) - \omega(\infty)$, and the steady-state error defined by $e_\omega(\infty)$. After determining the controller gains using the optimization algorithm, tests are conducted by applying disturbances to the system to evaluate control performance.

In this Case Study I, the inductance profile obtained through parametric regression is used, applying speed control to the SRM as illustrated in Fig. 7.1. In this scenario, the reference peak current is set at $I_{p_{ref}} = 10$ A, and the adopted switching angles, $\theta_{on} = 0°$ and $\theta_{off} = 30°$, follow the recommendations of the literature. A step signal is applied to the reference speed $\omega_{ref} = 155$ rad/s, with a load torque $C_c = 5$ N · m at $t = 25$ s. The hybrid optimization algorithm is used to adjust the gains of the PID controller [3]. The evaluation function considered is given by (7.1), resulting in the gains $k_p = 1.347$ [°·s/rad], $k_i = 8.040 \cdot 10^{-1}$ [°·s/rad], and $k_d = 6.409 \cdot 10^{-4}$ [°·s/rad]. Figure 7.2 presents the results obtained with the optimized PID controller acting on the excitation voltage with fixed switching angles [1].

The value of the evaluation function in the optimization process that determines the gains of the PID controller is $f_\omega = 1.97\%$. Figure 7.2b shows the maximum

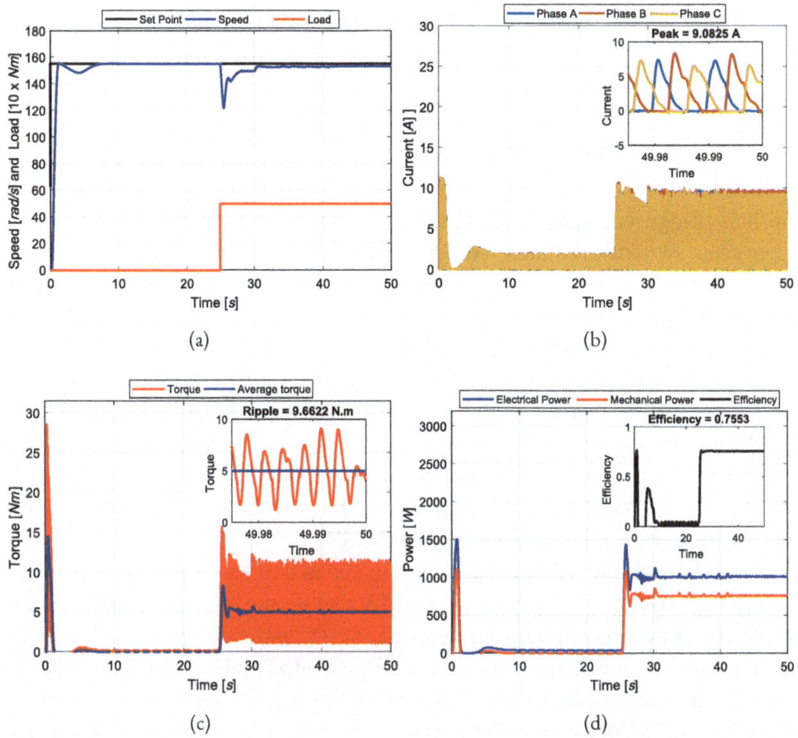

Fig. 7.2 PID controller operating through the excitation voltage with fixed switching angles: (**a**) speed response with $\omega_{ref} = 155$ rad/s, (**b**) phase current, (**c**) torque generated by the SRM, and (**d**) input electrical power, output mechanical power, and energy efficiency of the SRM

current during the start of the SRM, estimated at $I_p \approx 10$ A, and after the application of mechanical load, with $I_p \approx 9$ A. The average torque generated by the SRM is $C_m \approx 5\,\text{N}\cdot\text{m}$, while the torque ripple is $r_T \approx 9.6\,\text{N}\cdot\text{m}$, as illustrated in Fig. 7.2c. In Figure 7.2d, the average electrical power consumed is $P_e \approx 1000$ W, and the average power converted into mechanical power by the SRM is $P_m \approx 755$ W. The excitation voltage applied by the PID controller in this case is $V_{exc} \approx 235$ V.

The energy efficiency of the SRM in this case study is $\eta \approx 0.755$, as shown in Fig. 7.2d, under the application of mechanical load on the shaft. The study considers conventional operation with control applied to the excitation voltage and fixed switching angles within a 30° window. In this scenario, the system exhibits a response to speed control with correction after mechanical load application in less than 7 s, and energy efficiency within standard levels, considering that the machine used was manufactured manually [2].

7.2 Speed Control with Fixed Excitation Voltage

To achieve this proposal, different speed control topologies for the SRM are analyzed, allowing a comparative study of control conditions, energy efficiency, current ripples, mechanical torque, and electrical and mechanical power, among other factors. In this case, speed control is performed through the turn-off angle of the half-bridge converter switches $S_{\theta_{off}}$, while the firing angle α of the AC-DC converter switches remains fixed, ensuring a constant excitation voltage V_{exc}. Figure 7.3 illustrates the control topology acting on $S_{\theta_{off}}$, where the speed control is highlighted by the blue line, directly influencing the reference current of the coils I_{ref}, represented in the loop by the red line [1].

The PID speed controller acts directly on the reference current I_{ref} of the hysteresis controller. During switching, the coil current must be controlled using the soft-chopping technique, allowing a reduction in current ripple levels. This control technique is expected to minimize vibrations and torque ripples at high speeds. Studies indicate that controlling through the turn-off angle of the coils may negatively impact the energy efficiency of the SRM.

In Case Study II, the SRM is subjected to speed control as proposed in Fig. 7.3. The excitation voltage is set at $V_{exc} = 385$ V, while the switching angles adopted are those recommended in the literature: $\theta_{on} = 0°$ and $\theta_{off} = 30°$. A step signal is applied with the following parameters: $\omega_{ref} = 155$ rad/s and $C_c = 5$ N · m in $t = 25$ s. The hybrid optimization algorithm, with the evaluation function given by (7.1), is used to determine the gains of the PID controller, resulting in $k_p = 1.504 \cdot 10^{-1}$ [A · s/rad], $k_i = 6.710 \cdot 10^{-2}$ [A · s/rad], and $k_d = 1.215 \cdot 10^{-4}$ [A · s/rad]. Figure 7.4 presents the results obtained with the optimized PID controller, which acts on the coil current through the switching angles and the fixed excitation voltage [1].

The evaluation function in the optimization process that determines the gains of the PID controller is $f_\omega = 4.51\%$. Figure 7.4b shows the maximum current

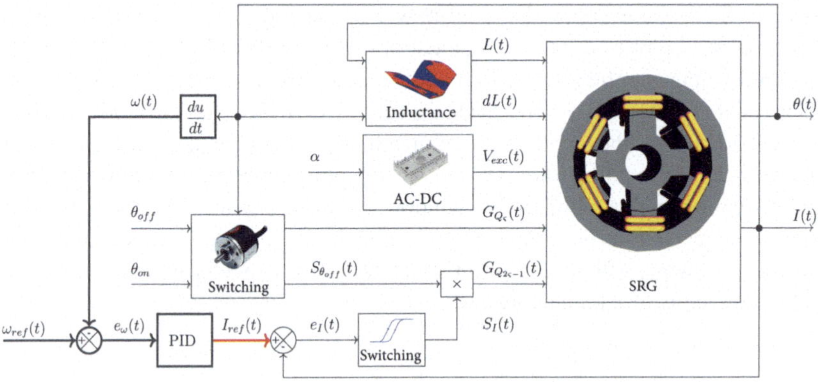

Fig. 7.3 Speed control system based on switching angles with fixed excitation voltage

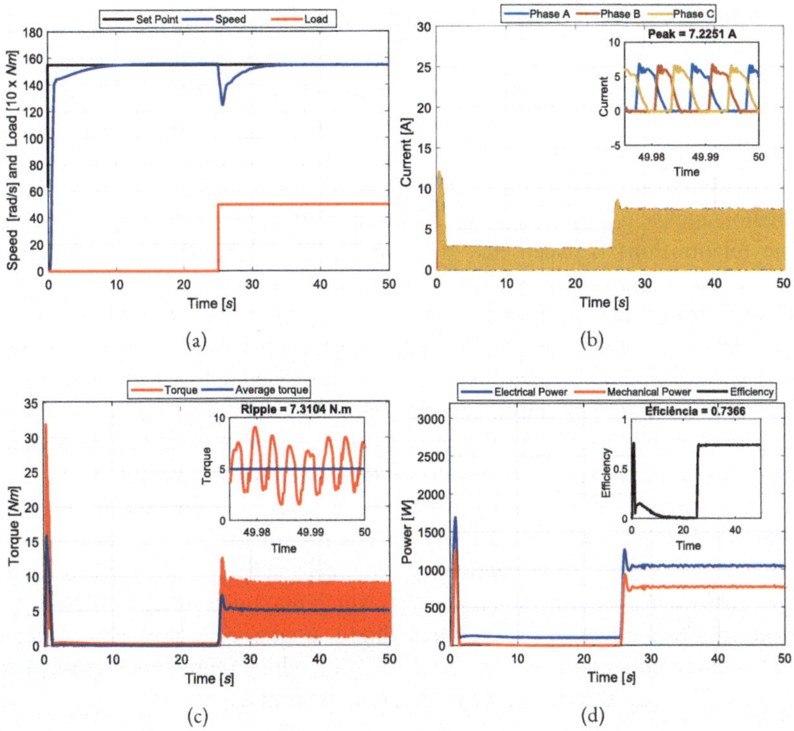

Fig. 7.4 Optimization of the PID controller acting on switching angles with fixed excitation voltage: (**a**) speed response with $\omega_{ref} = 155$ rad/s, (**b**) phase current, (**c**) torque generated by the SRM, and (**d**) input electrical power, output mechanical power, and energy efficiency of the SRM

during the start of the SRM, approximately 10 A, and after the application of the mechanical load, with $I_p \approx 7$ A. The average torque generated by the SRM is $C_m \approx 5\,\mathrm{N \cdot m}$, while the torque ripple is $r_T \approx 7.31\,\mathrm{N \cdot m}$, as illustrated in Fig. 7.4c. In Fig. 7.4d, the average electrical power consumed is $P_e \approx 1050$ W, and the average mechanical power delivered by the SRM is $P_m \approx 775$ W [2].

The energy efficiency of the SRM in this Case Study II is $\eta \approx 0.736$, as shown in Fig. 7.4d, under the application of mechanical load on the shaft. The study considers conventional operation, with control acting on the reference current through coil switching, using a fixed conduction window of 30°. In this scenario, the system demonstrates a correction response to the speed control after mechanical load application in less than 8 s and an energy efficiency 2.6% lower than in the previous case [2].

7.3 Speed Control with Dynamic Switching Angles

To achieve higher performance in speed control and energy efficiency, the excitation voltage control is employed with unconventional adjustments. Controlling the excitation voltage in the switched reluctance machine results in improved energy efficiency. Furthermore, studies indicate that for each value of ω, there are coil conduction angles, θ_{on} and θ_{off}, which maximize efficiency [1].

The relationships between θ_{on}, θ_{off}, and ω that optimize energy efficiency depend on the design characteristics of the SRM. However, analytical solutions for these relationships are not found in the literature. Therefore, this study proposes an approach to maximize energy efficiency η and ensure better control performance within a specific operating range. This model is illustrated in Fig. 7.5, where the blue line represents the speed control, and the control signal is depicted by the red line. Furthermore, in Fig. 7.5, the gray blocks represent the proposed system for dynamic adjustment of both the speed controller and the switching angles [2].

A reference speed range $\omega_{ref_{\min}} \leq \omega_{ref} \leq \omega_{ref_{\max}}$ is applied to the model, considering that $\omega_{ref}(\infty) \cong \omega(\infty)$ under control conditions. For each ω_{ref}, the ranges $\theta_{on_{\min}} \leq \theta_{on} \leq \theta_{on_{\max}}$ and $\theta_{off_{\min}} \leq \theta_{off} \leq \theta_{off_{\max}}$ are verified. Consequently, datasets $\theta_{on}(\omega_{ref})$ and $\theta_{off}(\omega_{ref})$, which maximize energy efficiency, are constructed. The parametric regression method is used to determine the coefficients of the Legendre polynomial $y_0, y_1, y_2, \cdots, y_n$, which represents $\theta_{on}(\omega_{ref})$, and $z_0, z_1, z_2, \cdots, z_n$, which represents $\theta_{off}(\omega_{ref})$, as given by:

$$\theta_{on}(\omega_{ref}) = y_0 + y_1 \cdot \omega_{ref} + y_2 \cdot \omega_{ref}^2 + \cdots + y_n \cdot \omega_{ref}^n \quad (7.2)$$

$$\theta_{off}(\omega_{ref}) = z_0 + z_1 \cdot \omega_{ref} + z_2 \cdot \omega_{ref}^2 + \cdots + z_n \cdot \omega_{ref}^n \quad (7.3)$$

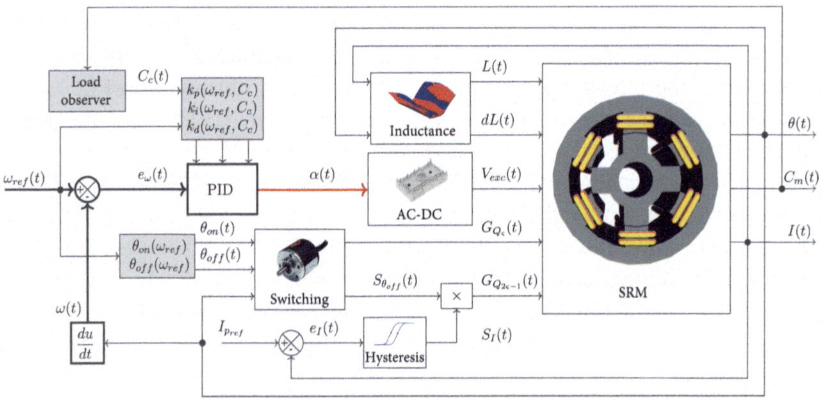

Fig. 7.5 Speed control system using excitation voltage with dynamic switching angles to improve energy efficiency

where n represents the number of coefficients required to optimize the curve fit. Since the switching angles in this case are dynamic, the system with input α and output ω also changes dynamically. For the PID controller to operate optimally throughout the range $\omega_{ref_{\min}} \leq \omega_{ref} \leq \omega_{ref_{\max}}$, the hybrid optimization algorithm determines the optimized gains k_p, k_i and k_d with the system under no load ($C_c = 0$) for each ω_{ref}, using the evaluation function defined in (7.1).

Subsequently, the hybrid optimization algorithm calculates the additional gains $k_{p_{Add}}$, $k_{i_{Add}}$, and $k_{d_{Add}}$ with the mechanical load system ($C_c > 0$), where C_c is dynamically calculated using the load observation system applied to the SRM shaft, as defined by expression (2.8). The parametric regression method is used to derive the mathematical expression that best fits the gain data in the entire range $\omega_{ref_{\min}} \leq \omega_{ref} \leq \omega_{ref_{\max}}$. The expressions in (7.4) to (7.6) correspond to the Legendre polynomial bases used to fit the curves associated with the gain sets $k_p(\omega_{ref}, C_c)$, $k_i(\omega_{ref}, C_c)$, and $k_d(\omega_{ref}, C_c)$.

$$\begin{aligned} k_p(\omega_{ref}, C_c) = &\underbrace{(A_0 + A_1 \cdot \omega_{ref} + A_2 \cdot \omega_{ref}^2 + \cdots + A_n \cdot \omega_{ref}^n)}_{\text{Ganho proporcional } k_p} + \\ &\underbrace{(a_0 + a_1 \cdot \omega_{ref} + a_2 \cdot \omega_{ref}^2 + \cdots + a_m \cdot \omega_{ref}^m) \cdot C_c}_{\text{Ganho proporcional adicional } k_{p_{Add}}} \end{aligned} \tag{7.4}$$

$$\begin{aligned} k_i(\omega_{ref}, C_c) = &\underbrace{(B_0 + B_1 \cdot \omega_{ref} + B_2 \cdot \omega_{ref}^2 + \cdots + B_n \cdot \omega_{ref}^n)}_{\text{Ganho integral } k_i} + \\ &\underbrace{(b_0 + b_1 \cdot \omega_{ref} + b_2 \cdot \omega_{ref}^2 + \cdots + b_m \cdot \omega_{ref}^m) \cdot C_c}_{\text{Ganho integral adicional } k_{i_{Add}}} \end{aligned} \tag{7.5}$$

$$\begin{aligned} k_d(\omega_{ref}, C_c) = &\underbrace{(C_0 + C_1 \cdot \omega_{ref} + C_2 \cdot \omega_{ref}^2 + \cdots + C_n \cdot \omega_{ref}^n)}_{\text{Ganho derivativo } k_d} + \\ &\underbrace{(c_0 + c_1 \cdot \omega_{ref} + c_2 \cdot \omega_{ref}^2 + \cdots + c_m \cdot \omega_{ref}^m) \cdot C_c}_{\text{Ganho derivativo adicional } k_{d_{Add}}} \end{aligned} \tag{7.6}$$

where n and m represent the number of coefficients required to optimize the curve fit. In this way, a system is structured to analytically calculate the values of $\theta_{on}(\omega)$ and $\theta_{off}(\omega)$ that maximize energy efficiency, as well as the values of k_p, k_i, and k_d that optimize speed control performance [2].

In this new Case Study III, based on the control loop illustrated in Fig. 7.5, a procedure is carried out to determine θ_{on} and θ_{off} that maximize SRM efficiency

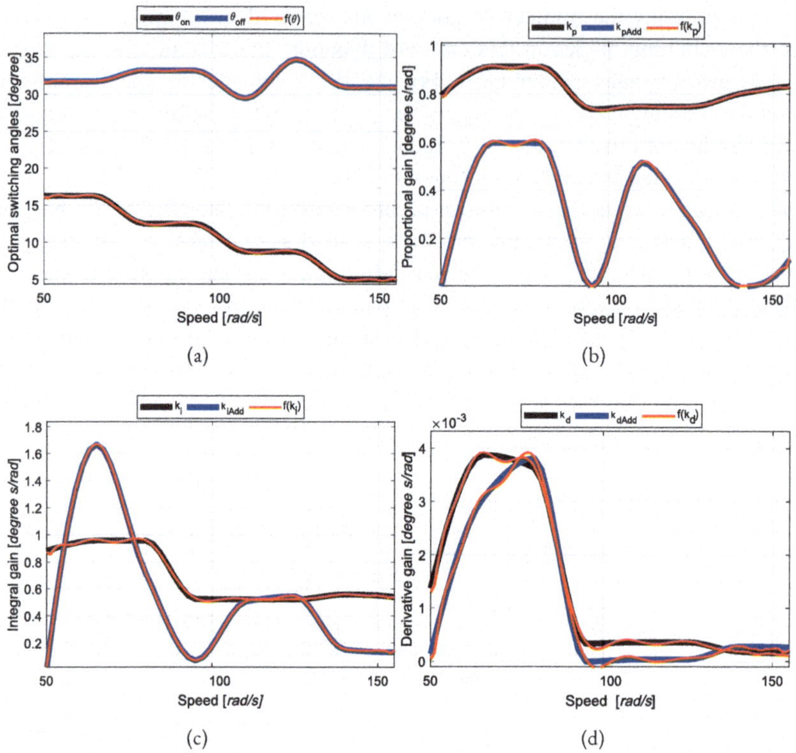

Fig. 7.6 Identification of analytical expressions representing the dynamics, as a function of speed, of the switching angles and PID controller gains: (**a**) switching angles of the SRM coils, (**b**) proportional action of the PID controller, (**c**) integral action of the PID controller, and (**d**) derivative action of the PID controller

for each ω_{ref} with $C_c = 5\,\text{N} \cdot \text{m}$. The test consists of varying the switching angles within the ranges $-5° \leq \theta_{on} \leq 15°$ and $-20° \leq \theta_{off} \leq 40°$, with increments of 2°, for each speed in the range $50\,\text{rad/s} \leq \omega_{ref} \leq 155\,\text{rad/s}$, with increments of 5 rad/s. For this analysis, the PID controller gains used are the same as those obtained by the hybrid optimization algorithm in Case Study I.

With the values of θ_{on} and θ_{off} that maximize the energy efficiency of each speed, the least squares method is applied to determine the coefficients of expressions (7.2) and (7.3), fitting the curves θ_{on} (black line) and θ_{off} (blue line) presented in Fig. 7.6a.

The tests are carried out on expressions (7.2) and (7.3) using the least squares method with $1 \leq n \leq 50$, in order to determine the value of n that provides the optimized expression. The expressions obtained using the proposed method are presented in (7.7) and (7.8), represented by the red curves in Fig. 7.6a [1].

$$\begin{aligned}\theta_{on}(\omega_{ref}) = &+9.390\cdot10^{7}-1.790\cdot10^{7}\cdot\omega_{ref}+1.590\cdot10^{6}\cdot\omega_{ref}^{2}-\\ &8.747\cdot10^{4}\cdot\omega_{ref}^{3}+3.336\cdot10^{3}\cdot\omega_{ref}^{4}-9.365\cdot10^{1}\cdot\omega_{ref}^{5}+\\ &2.003\cdot\omega_{ref}^{6}-3.337\cdot10^{-2}\cdot\omega_{ref}^{7}+4.382\cdot10^{-4}\cdot\omega_{ref}^{8}-\\ &4.564\cdot10^{-6}\cdot\omega_{ref}^{9}+3.769\cdot10^{-8}\cdot\omega_{ref}^{10}-2.445\cdot10^{-10}\cdot\omega_{ref}^{11}+\\ &1.246\cdot10^{-12}\cdot\omega_{ref}^{12}-4.825\cdot10^{-15}\cdot\omega_{ref}^{13}+1.376\cdot10^{-17}\cdot\omega_{ref}^{14}-\\ &2.727\cdot10^{-20}\cdot\omega_{ref}^{15}+3.352\cdot10^{-23}\cdot\omega_{ref}^{16}-1.923\cdot10^{-26}\cdot\omega_{ref}^{17}\end{aligned} \tag{7.7}$$

$$\begin{aligned}\theta_{on}(\omega_{ref}) = &+9.667\cdot10^{7}-1.859\cdot10^{7}\cdot\omega_{ref}+1.661\cdot10^{6}\cdot\omega_{ref}^{2}-\\ &9.170\cdot10^{4}\cdot\omega_{ref}^{3}+3.501\cdot10^{3}\cdot\omega_{ref}^{4}-9.814\cdot10^{1}\cdot\omega_{ref}^{5}+\\ &2.091\cdot\omega_{ref}^{6}-3.459\cdot10^{-2}\cdot\omega_{ref}^{7}+4.499\cdot10^{-4}\cdot\omega_{ref}^{8}-\\ &4.627\cdot10^{-6}\cdot\omega_{ref}^{9}+3.764\cdot10^{-8}\cdot\omega_{ref}^{10}-2.408\cdot10^{-10}\cdot\omega_{ref}^{11}\\ &+1.197\cdot10^{-12}\cdot\omega_{ref}^{12}-4.530\cdot10^{-15}\cdot\omega_{ref}^{13}+\\ &1.259\cdot10^{-17}\cdot\omega_{ref}^{14}-2.427\cdot10^{-20}\cdot\omega_{ref}^{15}+3.893\cdot10^{-23}\cdot\omega_{ref}^{16}-\\ &1.607\cdot10^{-26}\cdot\omega_{ref}^{17}\end{aligned} \tag{7.8}$$

With the expressions (7.7) and (7.8) known and integrated into the model, the PID controller with static gains operates in a non-optimized manner, considering the operating range of ω and the dynamic values of θ_{on} and θ_{off}. Thus, the static gains of the PID controller are optimized for each speed within the range 50 rad/s $\leq \omega_{ref} \leq$ 155 rad/s, using the hybrid optimization algorithm with the evaluation function defined in (7.1). Figure 7.6b–d present the results for the gains and the additional gains of the PID controller, allowing optimized adjustment throughout the range 50 rad/s $\leq \omega_{ref} \leq$ 155 rad/s [1].

With the gains and additional gains obtained, the least squares method is applied to determine the analytical expression that best fits the gains. The basis for the curve fitting expressions is defined in (7.4), (7.5), and (7.6). The tests are carried out using the least squares method for $1 \leq n \leq 50$ and $1 \leq m \leq 50$ to identify the values of n and m that provide optimized expressions. The resulting expressions are presented in (7.9), (7.10), and (7.11) and are represented by the red curves in Fig. 7.6b–d.

$$k_p(\omega_{ref}, C_c) = \big(+ 5.035 \cdot 10^6 - 9.475 \cdot 10^5 \cdot \omega_{ref} + 8.291 \cdot 10^4 \cdot \omega_{ref}^2 -$$
$$4.481 \cdot 10^3 \cdot \omega_{ref}^3 + 1.675 \cdot 10^2 \cdot \omega_{ref}^4 - 4.601 \cdot \omega_{ref}^5 +$$
$$9.660 \cdot 10^{-2} \cdot \omega_{ref}^6 - 1.558 \cdot 10^{-3} \cdot \omega_{ref}^7 + 1.987 \cdot 10^{-5} \cdot \omega_{ref}^8 -$$
$$2.006 \cdot 10^{-7} \cdot \omega_{ref}^9 + 1.602 \cdot 10^{-9} \cdot \omega_{ref}^{10} - 1.007 \cdot 10^{-11} \cdot \omega_{ref}^{11} +$$
$$4.925 \cdot 10^{-14} \cdot \omega_{ref}^{12} - 1.833 \cdot 10^{-16} \cdot \omega_{ref}^{13} + 5.021 \cdot 10^{-19} \cdot \omega_{ref}^{14} -$$
$$9.532 \cdot 10^{-22} \cdot \omega_{ref}^{15} + 1.120 \cdot 10^{-24} \cdot \omega_{ref}^{16} - 6.140 \cdot 10^{-28} \cdot \omega_{ref}^{17}) +$$
$$\big(- 5.710 \cdot 10^6 + 1.190 \cdot 10^6 \cdot \omega_{ref} - 1.150 \cdot 10^5 \cdot \omega_{ref}^2 +$$
$$6.853 \cdot 10^3 \cdot \omega_{ref}^3 - 2.816 \cdot 10^2 \cdot \omega_{ref}^4 + 8.481 \cdot \omega_{ref}^5 -$$
$$1.937 \cdot 10^{-1} \cdot \omega_{ref}^6 + 3.430 \cdot 10^{-3} \cdot \omega_{ref}^7 - 4.766 \cdot 10^{-5} \cdot \omega_{ref}^8 +$$
$$5.228 \cdot 10^{-7} \cdot \omega_{ref}^9 - 4.530 \cdot 10^{-9} \cdot \omega_{ref}^{10} + 3.083 \cdot 10^{-11} \cdot \omega_{ref}^{11} -$$
$$1.628 \cdot 10^{-13} \cdot \omega_{ref}^{12} + 6.539 \cdot 10^{-16} \cdot \omega_{ref}^{13} - 1.928 \cdot 10^{-18} \cdot \omega_{ref}^{14} +$$
$$3.935 \cdot 10^{-21} \cdot \omega_{ref}^{15} - 4.965 \cdot 10^{-24} \cdot \omega_{ref}^{16} + 2.917 \cdot 10^{-27} \cdot \omega_{ref}^{17}\big) \cdot C_c \tag{7.9}$$

The analytical expressions obtained are integrated into the computational model developed, as proposed in Fig. 7.5, where the SRM is subjected to speed control. In this case, the reference peak current is set at $I_{p_{ref}} = 10\,\text{A}$, and the switching angles are dynamically calculated according to (7.7) and (7.8). A step signal is applied to the reference speed $\omega_{ref} = 155$ rad/s, with an applied load torque of $C_c = 5\,\text{N} \cdot \text{m}$ at $t = 25$ s. The gains of the controller are dynamically calculated based on the expressions in (7.9) to (7.11). Figure 7.7 presents the results obtained from the proposed model [4].

$$k_i(\omega_{ref}, C_c) = \big(+ 1.237 \cdot 10^7 - 2.326 \cdot 10^6 \cdot \omega_{ref} + 2.035 \cdot 10^5 \cdot \omega_{ref}^2 -$$
$$1.100 \cdot 10^4 \cdot \omega_{ref}^3 + 1.130 \cdot 10^2 \cdot \omega_{ref}^4 - 4.601 \cdot 10^1 \cdot \omega_{ref}^5 +$$
$$2.361 \cdot 10^{-1} \cdot \omega_{ref}^6 - 3.830 \cdot 10^{-3} \cdot \omega_{ref}^7 + 4.890 \cdot 10^{-5} \cdot \omega_{ref}^8 -$$
$$4.940 \cdot 10^{-7} \cdot \omega_{ref}^9 + 3.950 \cdot 10^{-9} \cdot \omega_{ref}^{10} - 2.487 \cdot 10^{-11} \cdot \omega_{ref}^{11} +$$
$$1.217 \cdot 10^{-13} \cdot \omega_{ref}^{12} - 4.541 \cdot 10^{-16} \cdot \omega_{ref}^{13} + 1.245 \cdot 10^{-18} \cdot \omega_{ref}^{14} -$$
$$2.370 \cdot 10^{-21} \cdot \omega_{ref}^{15} + 2.793 \cdot 10^{-24} \cdot \omega_{ref}^{16} - 1.535 \cdot 10^{-27} \cdot \omega_{ref}^{17}\big)$$

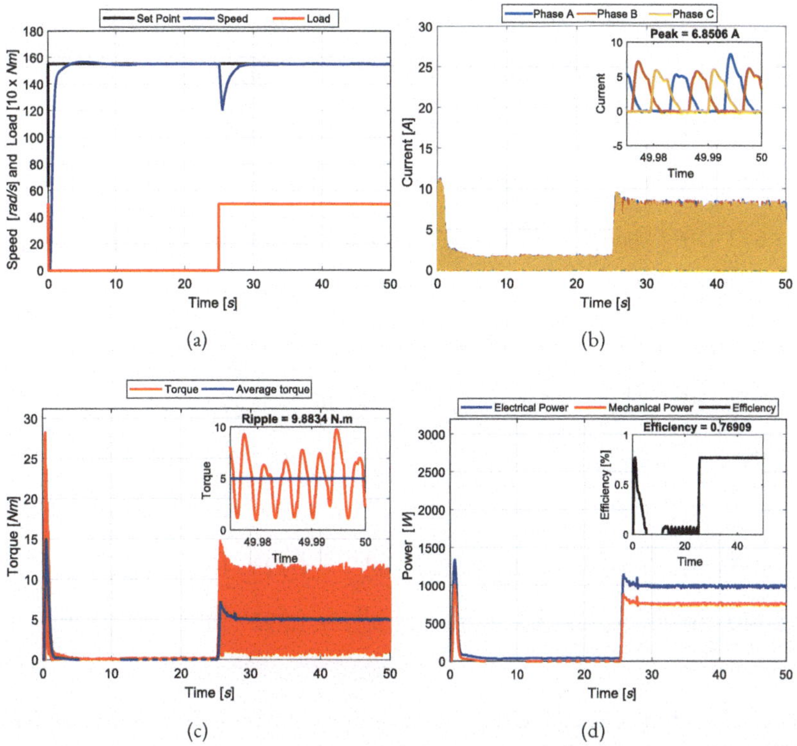

Fig. 7.7 PID controller operating on excitation voltage with dynamic PID adjustment and switching angles: (**a**) speed response with $\omega_{ref} = 155$ rad/s, (**b**) phase current, (**c**) torque generated by the SRM, and (**d**) input electrical power, output mechanical power, and energy efficiency of the SRM

$$
\begin{aligned}
&+ \big(-1.362 \cdot 10^{7} + 2.714 \cdot 10^{6} \cdot \omega_{ref} - 2.517 \cdot 10^{5} \cdot \omega_{ref}^{2} + \\
&1.442 \cdot 10^{4} \cdot \omega_{ref}^{3} - 5.716 \cdot 10^{2} \cdot \omega_{ref}^{4} + 1.663 \cdot 10^{1} \cdot \omega_{ref}^{5} - \\
&3.683 \cdot 10^{-1} \cdot \omega_{ref}^{6} + 6.331 \cdot 10^{-3} \cdot \omega_{ref}^{7} - 8.559 \cdot 10^{-5} \cdot \omega_{ref}^{8} + \\
&9.154 \cdot 10^{-7} \cdot \omega_{ref}^{9} - 7.746 \cdot 10^{-9} \cdot \omega_{ref}^{10} + 5.157 \cdot 10^{-11} \cdot \omega_{ref}^{11} - \\
&2.669 \cdot 10^{-13} \cdot \omega_{ref}^{12} + 1.051 \cdot 10^{-15} \cdot \omega_{ref}^{13} - 3.046 \cdot 10^{-18} \cdot \omega_{ref}^{14} + \\
&6.115 \cdot 10^{-21} \cdot \omega_{ref}^{15} - 7.599 \cdot 10^{-24} \cdot \omega_{ref}^{16} + 4.401 \cdot 10^{-27} \cdot \omega_{ref}^{17}\big) \cdot C_c
\end{aligned}
\tag{7.10}
$$

$$
\begin{aligned}
k_d(\omega_{ref}, C_c) = \big(&+8.671 \cdot 10^4 - 1.633 \cdot 10^4 \cdot \omega_{ref} + 1.431 \cdot 10^3 \cdot \omega_{ref}^2 - \\
&7.748 \cdot 10^1 \cdot \omega_{ref}^3 + 2.901 \cdot \omega_{ref}^4 - 7.980 \cdot 10^{-2} \cdot \omega_{ref}^5 + \\
&1.669 \cdot 10^{-3} \cdot \omega_{ref}^6 - 2.712 \cdot 10^{-5} \cdot \omega_{ref}^7 + 3.467 \cdot 10^{-7} \cdot \omega_{ref}^8 - \\
&3.507 \cdot 10^{-9} \cdot \omega_{ref}^9 + 2.808 \cdot 10^{-11} \cdot \omega_{ref}^{10} - 1.770 \cdot 10^{-13} \cdot \omega_{ref}^{11} + \\
&8.682 \cdot 10^{-16} \cdot \omega_{ref}^{12} - 3.242 \cdot 10^{-18} \cdot \omega_{ref}^{13} + 8.912 \cdot 10^{-21} \cdot \omega_{ref}^{14} - \\
&1.698 \cdot 10^{-23} \cdot \omega_{ref}^{15} + 2.005 \cdot 10^{-26} \cdot \omega_{ref}^{16} - 1.104 \cdot 10^{-29} \cdot \omega_{ref}^{17}\big) + \\
\big(&+1.089 \cdot 10^5 - 2.052 \cdot 10^4 \cdot \omega_{ref} + 1.798 \cdot 10^3 \cdot \omega_{ref}^2 - \\
&9.734 \cdot 10^1 \cdot \omega_{ref}^3 + 3.646 \cdot \omega_{ref}^4 - 1.003 \cdot 10^{-1} \cdot \omega_{ref}^5 + \\
&2.098 \cdot 10^{-3} \cdot \omega_{ref}^6 - 3.410 \cdot 10^{-5} \cdot \omega_{ref}^7 + 4.361 \cdot 10^{-7} \cdot \omega_{ref}^8 - \\
&4.413 \cdot 10^{-9} \cdot \omega_{ref}^9 + 3.535 \cdot 10^{-11} \cdot \omega_{ref}^{10} - 2.229 \cdot 10^{-13} \cdot \omega_{ref}^{11} + \\
&1.093 \cdot 10^{-15} \cdot \omega_{ref}^{12} - 4.085 \cdot 10^{-18} \cdot \omega_{ref}^{13} + 1.123 \cdot 10^{-20} \cdot \omega_{ref}^{14} - \\
&2.140 \cdot 10^{-23} \cdot \omega_{ref}^{15} + 2.527 \cdot 10^{-26} \cdot \omega_{ref}^{16} - 1.392 \cdot 10^{-29} \cdot \omega_{ref}^{17}\big) \cdot C_c
\end{aligned}
\tag{7.11}
$$

The value of the evaluation function determined by the expression (7.1), for the test where $\omega_{ref} = 155$ rad/s, is $f_\omega = 1.17\%$. Figure 7.7b shows the maximum current during the start of the SRM, approximately $I_p \approx 10$ A, and after the application of the mechanical load, with $I_p \approx 6.8$ A. The average torque generated by the SRM is $C_m \approx 5\text{N} \cdot \text{m}$, while the torque ripple is $r_T \approx 9.8\,\text{N} \cdot \text{m}$, as illustrated in Fig. 7.7c. In Fig. 7.7d, the average electrical power consumed is $P_e \approx 990$ W, and the average power converted into mechanical power by the SRM is $P_m \approx 761$ W. The excitation voltage applied by the PID controller in this case is $V_{exc} \approx 341$ V. As shown in Fig. 7.7d, the energy efficiency of the SRM for this case study is $\eta \approx 0.769$, with the application of mechanical load on the shaft [2].

The study considers a nonconventional drive with analytically assisted control operating on the excitation voltage and dynamic switching angles. In this scenario, the system demonstrates a correction–correction response to speed control after the application of approximately 4 s mechanical load and an energy efficiency 1.8% higher compared to the conventional drive in Case Study I and 4.3% higher than the drive presented in Case Study II. The tests are performed at a single SRM operating point, where $\omega = 155$ rad/s and $C_c = 5\,\text{N} \cdot \text{m}$ [4].

Table 7.1 Values of switching angles and speed controller gains for tests with a step applied at $\omega_{ref} = 155$ rad/s and $C_c = 5\,\text{N}\cdot\text{m}$

Case Study	θ_{on} [°]	θ_{off} [°]	k_p*	k_i*	k_d*
I	0	30	1.34	$8.04 \cdot 10^{-1}$	$6.40 \cdot 10^{-4}$
II	0	30	$1.50 \cdot 10^{-1}$	$6.71 \cdot 10^{-2}$	$1.21 \cdot 10^{-4}$
III	(7.8)	(7.8)	(7.9)	(7.10)	(7.11)

Table 7.2 Values of power, speed evaluation function, and energy efficiency for tests with a step applied at $\omega_{ref} = 155$ rad/s and $C_c = 5\,\text{N}\cdot\text{m}$

Case Study	P_e [W]	P_m [W]	f_ω [%]	η [%]
I	1000	755	1.97	0.755
II	1050	775	4.51	0.736
III	990	761	1.17	0.769

7.4 Establishing the Method Through Case Studies

The SRM drive and control tests are performed for a single operating point in Case Study I, Case Study II, and Case Study III, where the speed and load torque result in a mechanical power of approximately $P_m \approx 775$ W. The results obtained are presented in Tables 7.1 and 7.2. In addition, tests are performed throughout the operating range 50 rad/s $\leq \omega_{ref} \leq$ 155 rad/s with a load applied to the SRM shaft in the range $0.5\,\text{N}\cdot\text{m} \leq C_c \leq 5\,\text{N}\cdot\text{m}$.

In Tables 7.1[1] and 7.2, Case Study III shows the lowest evaluation function value for speed control and the lowest electrical power consumption P_e, resulting in the highest energy efficiency η [4].

Conventional drives with the PID controller operating at excitation voltage (Case Study I) and on coil switching (Case Study II) are optimized under the following conditions: (i) fixed speed $\omega_{ref} = 155$ rad/s and (ii) fixed load torque $C_c = 5\,\text{N}\cdot\text{m}$. Since the PID controller is a linear system with fixed gains, it is expected that the further the operating point is from the point of optimization, the lower the control response quality. Figure 7.8a presents the performance of 'the speed controller, evaluated using expression (7.1), for the three proposed drives [4].

The same range of ω_{ref} used to derive the expressions (7.7) to (7.11) is considered. In Case Study III, the SRM drive and control system behaves dynamically in the ω_{ref} range. Under extrapolation conditions, Case Study III behaves similarly to Case Study I, with fixed switching angles and controller gains. In Fig. 7.8a, it can be observed that Case Study III exhibits superior speed control performance compared to Case Study I and Case Study II across the entire operating range, with a 145% lower value compared to Case Study I and 113% lower compared to Case Study II.

[1] Unit of measurement: [° · s/rad] for Case Study I and Case Study III, and [A · s/rad] for Case Study II.

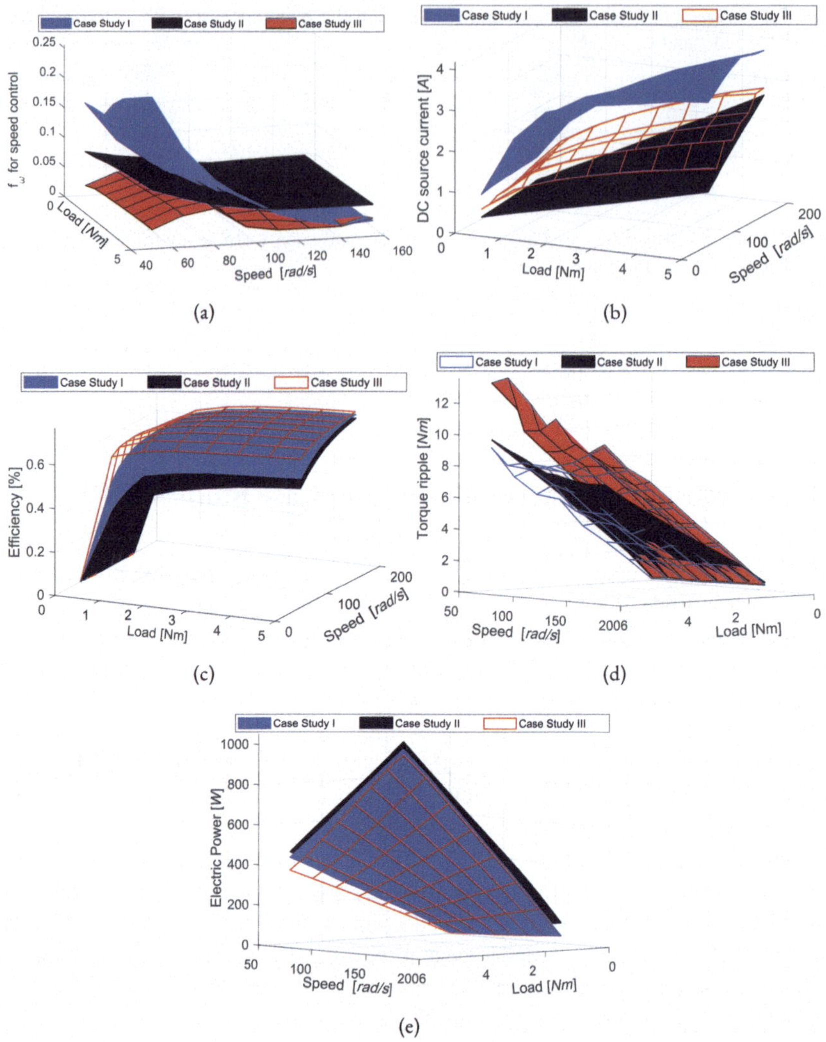

Fig. 7.8 Test in Case Study I, Case Study II, and Case Study III with step inputs applied in the ranges 50 rad/s $\leq \omega_{ref} \leq$ 150 rad/s and 0.5 N · m $\leq C_c \leq$ 5 N · m: (**a**) evaluation function for the PID controller, (**b**) energy efficiency, (**c**) DC bus current, (**d**) torque ripple of the SRM, and (**e**) electrical power consumed by the SRM

Considering the same speed and torque, the maximum difference in the evaluation function values for Case Study III compared to Case Study I is 699%, and 208% compared to Case Study II. The lower the value of the evaluation function, the more efficient the control. It is observed that the difference in the values of the evaluation function using the model proposed in Case Study III is greater under conditions where ω and C_c are further from the operating point at which Case Study I and

Table 7.3 Values of torque ripple, excitation current, and power under step inputs applied in the range 50 rad/s $\leq \omega_{ref} \leq$ 155 rad/s and 0.5 N · m $\leq C_c \leq$ 5 N · m

Case Study	$\overline{r_T}$ [N · m]	$\overline{I_{exc}}$ [A]	$\overline{P_e}$ [W]	$\overline{P_{mec}}$ [W]
I	4.53	2.82	375	256
II	4.35	1.14	418	256
III	5.69	1.69	352	255

Table 7.4 Evaluation function and energy efficiency values under step inputs applied in the range 50 rad/s $\leq \omega_{ref} \leq$ 155 rad/s and 0.5 N · m $\leq C_c \leq$ 5 N · m

Case Study	$f_{\omega_{max}}$ [%]	$f_{\omega_{min}}$ [%]	$\overline{f_\omega}$ [%]	η_{max} [%]	η_{min} [%]	$\overline{\eta}$ [%]
I	25.49	1.53	6.90	75.53	47.75	55.30
II	9.83	4.30	5.92	73.65	33.96	47.57
III	6.17	1.20	2.77	76.90	59.26	65.44

Case Study II were optimized (ω = 155 rad/s and C_c = 5 N · m). The dynamic gain PID control system achieves lower f_ω values throughout the operating range analyzed [2].

Regarding energy efficiency, as shown in Fig. 7.8b, throughout the operating range, Case Study III demonstrates superior performance, with 6.80% higher efficiency compared to Case Study I and 18.16% higher compared to Case Study II. Taking into account the same speed and torque, the largest difference in the evaluation function values for Case Study III compared to Case Study I is 23.65%, and 40.55% compared to Case Study II. It is observed that the difference in energy efficiency for the model proposed in Case Study III, compared to the other cases, is more pronounced at lower speeds and shows higher values throughout the operating range.

The DC current supplied to the SRM through the DC bus is shown in Fig. 7.8c. The model proposed in Case Study III consumes less current compared to Case Study I. The torque ripple is shown in Fig. 7.8d, where Case Study I exhibits lower ripple at higher speeds. Fig. 7.8e illustrates the electrical power consumption of the SRM, where, throughout the entire analysis range, Case Study III shows the lowest consumption while delivering the same mechanical power as the other case studies. These results are detailed in Tables 7.3 and 7.4.

In Tables 7.3 and 7.4, Case Study II shows the lowest mean torque ripple $\overline{r_T}$ and the lowest average excitation current consumption $\overline{I_{exc}}$. On the other hand, Case Study III exhibits the best values for $\overline{P_e}$, $f_{\omega_{max}}$, $f_{\omega_{min}}$, $\overline{f_\omega}$, η_{max}, η_{min}, and $\overline{\eta}$.

The test with variations in the speed controller setpoint and different mechanical load values applied to the SRM shaft is performed and illustrated in Fig. 7.9. This test has a simulation time of t = 408 s, with ω_{ref} = [140, 60, 125, 65] rad/s at intervals of t = 100 s, and C_c = 5 N · m applied to t = 20 s during each change in ω_{ref}. Figure 7.9a presents the speed response of the PID controller for Case Study I, Case Study II, and Case Study III. The best performance is observed on the red line, corresponding to Case Study III.

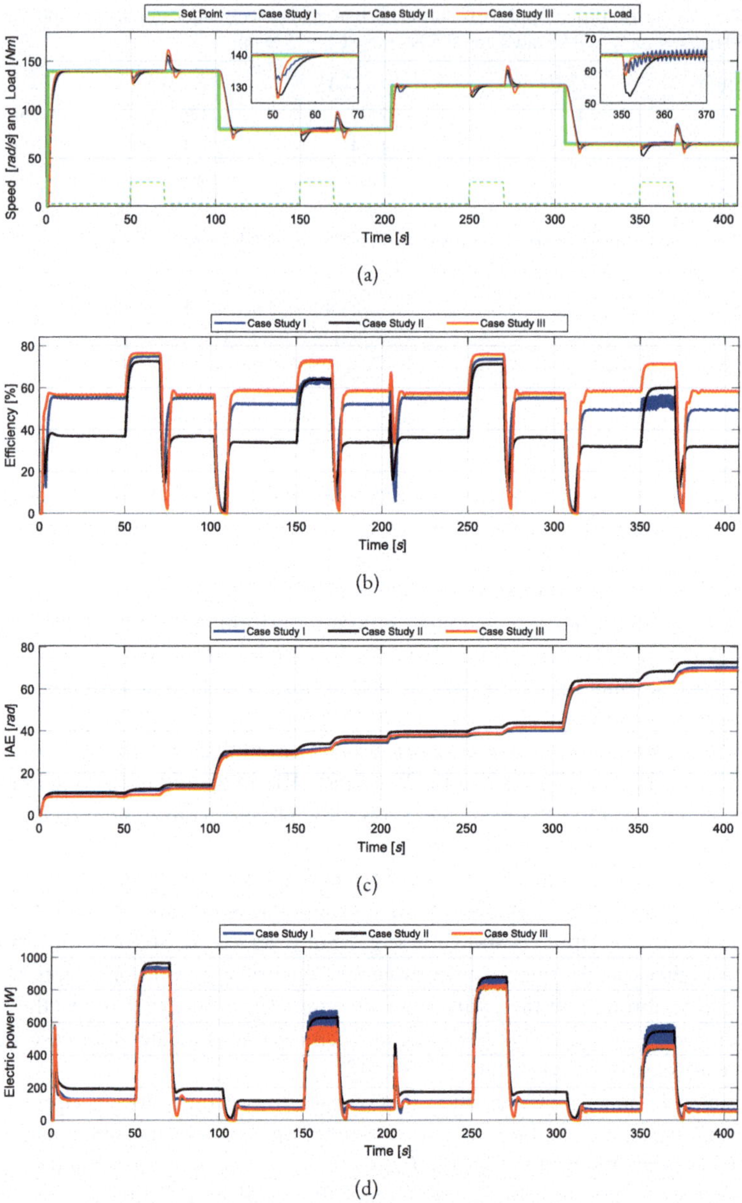

Fig. 7.9 Test with variation in ω_{ref} and C_c: (**a**) SRM speed and load torque ($C_c \cdot 5$), (**b**) energy efficiency, (**c**) accumulated absolute error, and (**d**) electrical power consumed by the SRM

Table 7.5 Power values for $\omega_{ref} = [140, 80, 125, 65]$ rad/s at intervals of $t = 100$ s, and $C_c = 5\,\text{N}\cdot\text{m}$ applied for $t = 20$ s during each change in ω_{ref}

Case Study	$P_{e_{\max}}$ $[W]$	$P_{e_{\min}}$ $[W]$	$\overline{P_e}$ $[W]$	$P_{m_{\max}}$ $[W]$	$P_{m_{\min}}$ $[W]$	$\overline{P_m}$ $[W]$
I	944	66.98	216.99	708	32	141
II	967	406.90	260.41	702	52	140
III	920	56.65	242	704	52	141

Table 7.6 Accumulated error and energy efficiency values for $\omega_{ref} = [140, 80, 125, 65]$ rad/s at intervals of $t = 100$ s, and $C_c = 5\,\text{N}\cdot\text{m}$ applied for $t = 20$ s during each change in ω_{ref}

Case Study	IAE [rad]	$\eta_{\max}$ [%]	$\eta_{\min}$ [%]	$\overline{\eta}$ [%]
I	762	74.96	49.57	51.45
II	744	72.64	31.95	38, 84, 56
III	726	76.55	58.67	56.50

In Fig. 7.9a, the load torque, represented by the dashed green line, is multiplied by 5 to improve visibility. Figure 7.9b shows the energy efficiency values of the SRM under the test conditions for Case Study I, Case Study II, and Case Study III. Case Study III exhibits the highest efficiency values compared to the other case studies, particularly when the SRM operates at low speeds. Figure 7.9c presents the absolute error values accumulated in $t = 408$ s, where Case Study III shows 2.38% less accumulated error compared to Case Study I and 4.93% less compared to Case Study II.

The electrical power consumption throughout the interval t is shown in Fig. 7.9d, where the electrical power accumulated per second is $P_e = 88.98$ kW for Case Study I, $P_e = 106.39$ kW for Case Study II, and $P_e = 83.33$ kW for Case Study III. Case Study III consumes 6.78% less power compared to Case Study I and 27.67% less compared to Case Study II, demonstrating better speed control performance, especially during moments of application of mechanical load. These results are presented in Tables 7.5 and 7.6.

In Tables 7.5 and 7.6, Case Study III shows the best values for $P_{e_{\max}}$, $P_{e_{\min}}$, $\overline{P_e}$, IAE, $\eta_{\max}$, $\eta_{\min}$, and $\overline{\eta}$ compared to Case Study I and Case Study II. The results presented enable a comparative analysis of the drive and control methods for the switched reluctance motor, all employing the parametric regression method in the simulation model. Furthermore, the study demonstrates an optimized method that ensures stability of the speed control and energy efficiency of the SRM within the defined operating range. Based on the simulation results, it becomes feasible to conduct an experimental validation of the proposal on a test bench, considering Case Study I and Case Study III.

7.5 Validation of the Driving and Control Techniques on a Test Bench

Case Study I and Case Study III demonstrate the best performance in speed control and energy efficiency during simulations and are therefore selected for implementation on the test bench. This choice enables the validation of the proposed methodology, both in the use of the inductance surface obtained through the parametric regression method and in the speed-assisted control methodology using excitation voltage with dynamic switching angles [2].

To validate the proposed methodology, new simulations are performed, and the findings are compared with those obtained from the test bench using the normalized difference between the simulated and experimental data (deviation). The case studies are configured to operate within the speed range 50 rad/s $\leq \omega_{ref} \leq$ 155 rad/s and load torque range $0.5\,\text{N}\cdot\text{m} \leq C_c \leq 5\,\text{N}\cdot\text{m}$. In test bench experiments, the load torque C_c is generated based on the electric current I_L applied to the electromagnetic brake (Fig. 5.11).

During the validation process, an electric current of $I_L = 20\,\text{A}$ is used, and the load torque C_c produced by the electromagnetic brake is applied as an input parameter in the simulations to ensure that the model and the system operate under similar conditions. Figure 7.10 presents a comparison between the experimental data and the simulations for the optimized PID controller, acting on the excitation voltage with fixed switching angles [2].

The load torque generated by the electromagnetic brake is $C_c = 1.6\,\text{N}\cdot\text{m}$ at $t = 25\,\text{s}$, with the load torque, represented by the green line, multiplied by *ten* to improve visibility in Fig. 7.10a. The speed is controlled at the reference value of 155 rad/s with reduced overshoot. The simulated integral of the absolute error (IAE_s) is approximately 1.95%, while the experimental integral (IAE_e) is approximately 5.29%, resulting in a deviation of 1.712 between the experimental and simulated values. Figure 7.10b shows the simulated electrical power (P_{e_s}) of approximately 370 W and the experimental electrical power (P_{e_e}) of approximately 360 W, with a deviation of 0.027. Figure 7.10c presents the simulated mechanical power (P_{m_s}) of approximately 255 W and the experimental mechanical power (P_{m_e}) of approximately 248 W, with a deviation of 0.031 [4].

The simulated energy efficiency (η_s) is approximately 0.691, while the experimental efficiency (η_e) is approximately 0.688, with a deviation of 0.004, as shown in Fig. 7.10d. The study considers conventional operation with control acting on the excitation voltage and fixed switching angles within a 30° window. In this context, the bench system exhibits a control response similar to the simulated model, correcting the speed error and achieving energy efficiency levels consistent with the methodology proposed in Case Study I. To analyze the system at an operating point different from that used in the optimization process, a step signal is applied to the speed reference $\omega_{ref} = 95$ rad/s, with a load torque generated by the electromagnetic brake of $C_c = 1.53\,\text{N}\cdot\text{m}$ at $t = 25\,\text{s}$. The experimental results are compared with the simulations and presented in Fig. 7.11 [4].

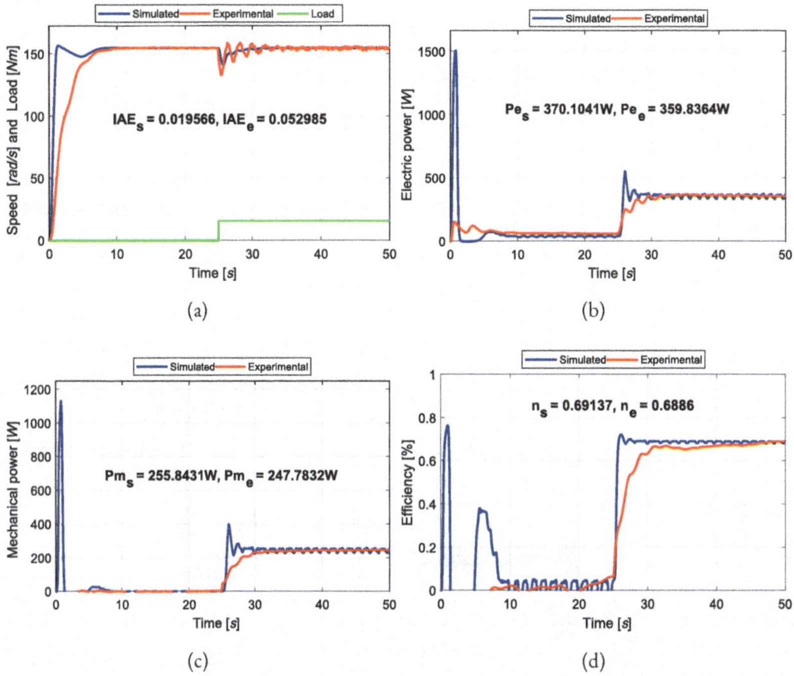

Fig. 7.10 Operation with the PID controller acting through the excitation voltage with fixed switching angles at $\omega_{ref} = 155$ rad/s, experimental and simulated: (**a**) speed response and load torque ($C_c \cdot 10$), (**b**) input electrical power, (**c**) output mechanical power, and (**d**) energy efficiency

The speed is controlled with a reference of 95 rad/s, with higher overshoot levels compared to the previous test, as the system operates outside the optimized point in this case. The simulated integral of the absolute error (IAE_s) is approximately 4.39%, while the experimental integral (IAE_e) is approximately 7.54%, with a deviation of 0.717 between the simulated and experimental values. Figure 7.11b shows the simulated electrical power (P_{e_s}) of approximately 213 W and the experimental electrical power (P_{e_e}) of approximately 233 W, with a deviation of 0.095. The simulated mechanical power (P_{m_s}) is approximately 138 W, while the experimental mechanical power (P_{m_e}) is approximately 149 W, with a deviation of 0.087, as shown in Fig. 7.11c [2].

The simulated energy efficiency (η_s) of the SRM is approximately 0.642, while the experimental efficiency (η_e) is approximately 0.651, as presented in Fig. 7.11d, with a deviation of 0.014. It is observed that the speed control performance decreases when operating far from the optimized point, due to the system's high nonlinearity and the fact that the PID controller with fixed gains operates in a linear manner [1].

To improve the performance of speed control at different operating points, dynamic gains are applied to the PID controller. Thus, Case Study III is used to

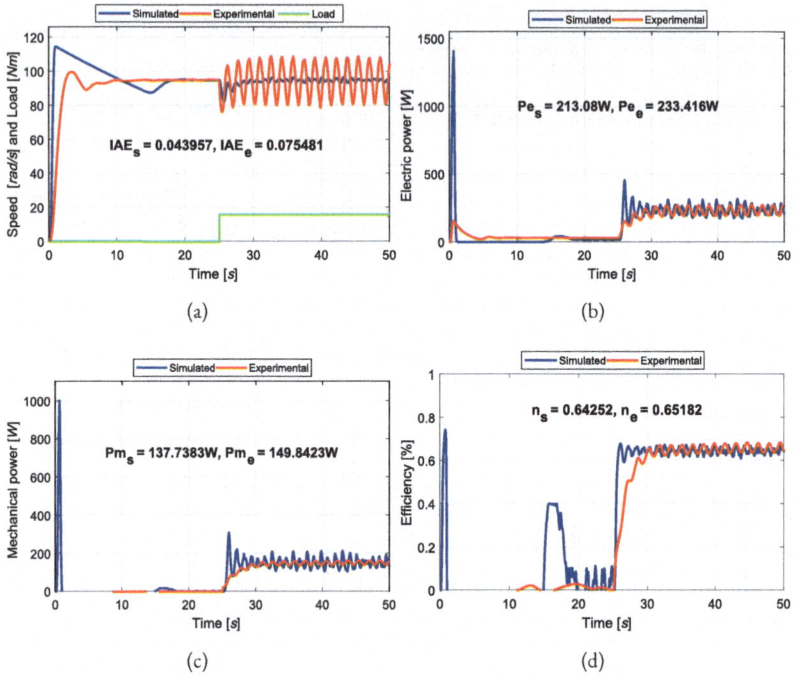

Fig. 7.11 Operation with the PID controller acting through the excitation voltage with fixed switching angles at $\omega_{ref} = 95$ rad/s, experimental and simulated: (**a**) speed response and load torque ($C_c \cdot 10$), (**b**) input electrical power, (**c**) output mechanical power, and (**d**) energy efficiency

achieve better speed control and energy efficiency performance within a specific operating range. Figure 7.12 presents the comparison between experiments and simulations for the assisted PID controller that acts on the excitation voltage with dynamic switching angles.

The load torque generated by the electromagnetic brake is $C_c = 1.6\,\mathrm{N \cdot m}$ at $t = 25$ s. The speed is controlled with a reference of 155 rad/s, with no oscillations after applying the mechanical load. The simulated integral of the absolute error (IAE_s) is approximately 2.10%, while the experimental integral (IAE_e) is approximately 5.39%, with a deviation of 1.566 between the simulated and experimental values. Figure 7.12b shows the simulated electrical power (P_{e_s}) of approximately 360 W and the experimental electrical power (P_{e_e}) of approximately 357 W, with a deviation of 0.008. The simulated mechanical power (P_{m_s}) is approximately 254 W, while the experimental mechanical power (P_{m_e}) is approximately 248 W, with a deviation of 0.021, as shown in Fig. 7.12c [4].

The simulated energy efficiency (η_s) of the SRM is approximately 0.704, while the experimental efficiency (η_e) is approximately 0.696, as presented in Fig. 7.12d, with a deviation of 0.012. The study considers operation with dynamic control acting on the excitation voltage and dynamic switching angles. To analyze another

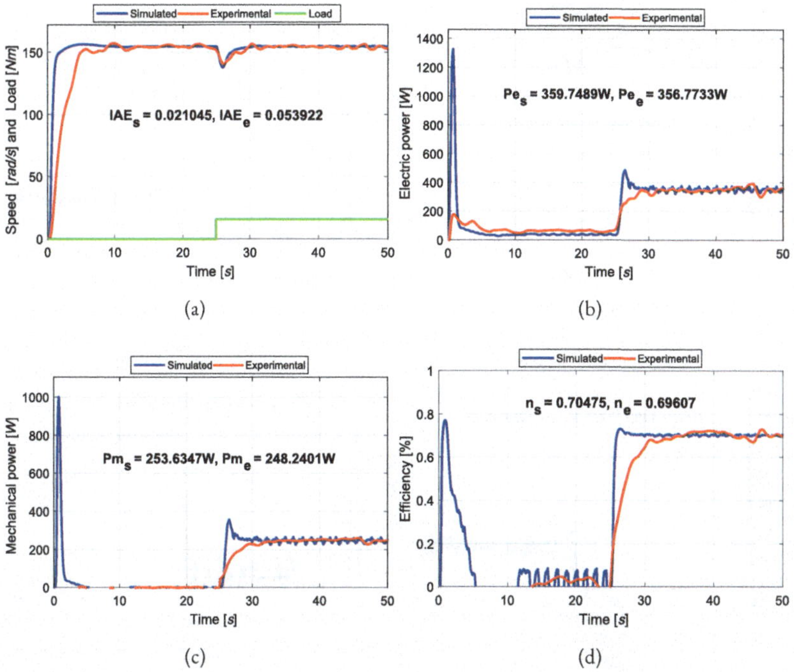

Fig. 7.12 Operation with the PID controller acting through the excitation voltage with PID adjustment and dynamic switching angles at $\omega_{ref} = 155$ rad/s, experimental and simulated: (**a**) speed response and load torque ($C_c \cdot 10$), (**b**) input electrical power, (**c**) output mechanical power, and (**d**) energy efficiency

operating point, a step signal is applied to the speed reference $\omega_{ref} = 95$ rad/s, with a load torque generated by the electromagnetic brake of $C_c = 1.53\,\text{N} \cdot \text{m}$ at $t = 25$ s. Figure 7.13 illustrates the operation with the PID controller acting through the excitation voltage.

The simulated integral of absolute error (IAE_s) is approximately 2.15%, while the experimental integral (IAE_e) is approximately 4.95%, with a deviation of 1.302 between the simulated and experimental values. Figure 7.13b shows the simulated electrical power (P_{e_s}) of approximately 203 W and the experimental electrical power (P_{e_e}) of approximately 222 W, with a deviation of 0.097. The simulated mechanical power (P_{m_s}) is approximately 140 W, while the experimental mechanical power (P_{m_e}) is approximately 152 W, with a deviation of 0.084, as shown in Fig. 7.13c. The simulated energy efficiency (η_s) of the SRM is approximately 0.692, while the experimental efficiency (η_e) is approximately 0.683, as presented in Fig. 7.13a, with a deviation of 0.012.

The speed is controlled at the reference value of 95 rad/s, with no oscillations observed after the application of the mechanical load. However, oscillations occur when the system operates under no-load conditions ($C_c = 0\,\text{N} \cdot \text{m}$). This behavior

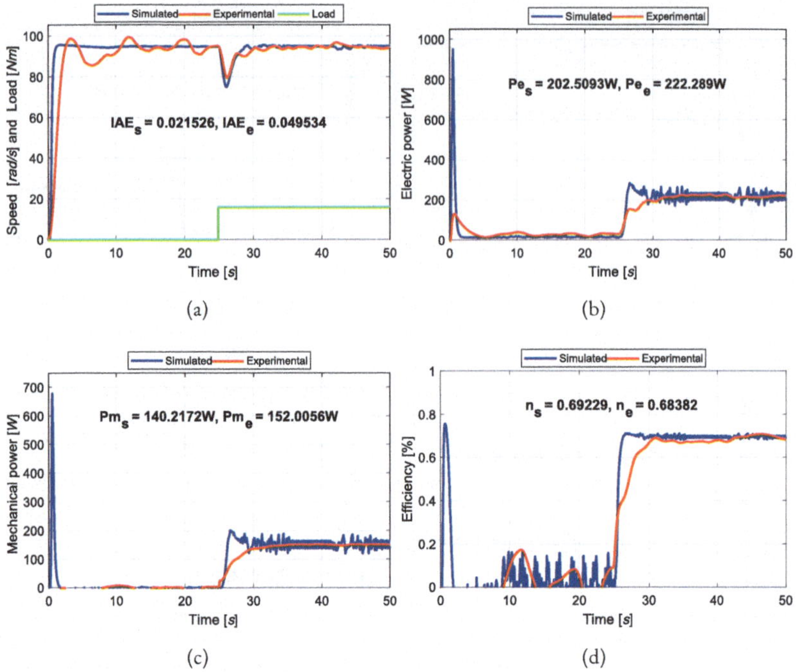

Fig. 7.13 Operation with the PID controller acting through the excitation voltage with PID adjustment and dynamic switching angles at $\omega_{ref} = 95$ rad/s, experimental and simulated: (**a**) speed response and load torque ($C_c \cdot 10$), (**b**) input electrical power, (**c**) output mechanical power, and (**d**) energy efficiency

arises due to control conditions, as lower ω_{ref} values lead to narrower conduction windows (the distance between θ_{on} and θ_{off} points in Fig. 7.6a) for the *half-bridge* converter switches. In this case, the system exhibits high nonlinearity, considering that the test bench introduces inherent and emergent variables that are typically not accounted for in the computational model.

In the validation of the techniques applied in Case Study I and Case Study III on the test bench, the experimental results show behavior similar to the computational model. The comparison is performed using the normalized difference between the simulated and experimental values (deviation). Tables 7.7 and 7.8 present the comparison between the simulation and test bench results for step tests at $\omega_{ref} = 155$ rad/s and $\omega_{ref} = 95$ rad/s, thus validating the model [4].

An analysis is performed in both the computational model and the test bench, applying speed steps within the range of 65 rad/s $\leq \omega_{ref} \leq$ 155 rad/s, in increments of 15 rad/s. The load torques generated by the electromagnetic brake, through the current I_L, are $C_c = [2.31,\ 1.61,\ 1.60,\ 1.53,\ 1.53,\ 1.59,\ 1.6]\,\mathrm{N \cdot m}$, applied at intervals of $t = 26$ s. The values obtained are presented in Fig. 7.14, where the blue and red lines represent experimental values, while the blue and red points correspond to simulated values [4].

Table 7.7 Tests performed in simulation and on the test bench for Case Study I and Case Study III with a step applied at $\omega_{ref} = 155$ rad/s

Case Study	IAE_s [%]	IAE_e [%]	P_{e_s} [W]	P_{e_e} [W]	P_{m_s} [W]	P_{m_e} [W]	η_s [%]	η_e [%]
I	1.95	5.29	370.10	359.83	255.84	247.78	69.13	68.86
III	2.10	5.39	359.74	356.77	253.63	248.24	70.47	69.60
Deviation I	1.712		0.027		0.031		0.004	
Deviation III	1.566		0.008		0.021		0.012	

Table 7.8 Tests performed in simulation and on the test bench for Case Study I and Case Study III with a step applied at $\omega_{ref} = 95$ rad/s

Case Study	IAE_s [%]	IAE_e [%]	P_{e_s} [W]	P_{e_e} [W]	P_{m_s} [W]	P_{m_e} [W]	η_s [%]	η_e [%]
I	4.39	7.54	213.08	233.41	137.73	149.84	64.25	65.18
III	2.15	4.95	202.50	222.28	140.21	152.00	69.22	68.38
Deviation I	0.717		0.095		0.087		0.014	
Deviation III	1.302		0.097		0.084		0.012	

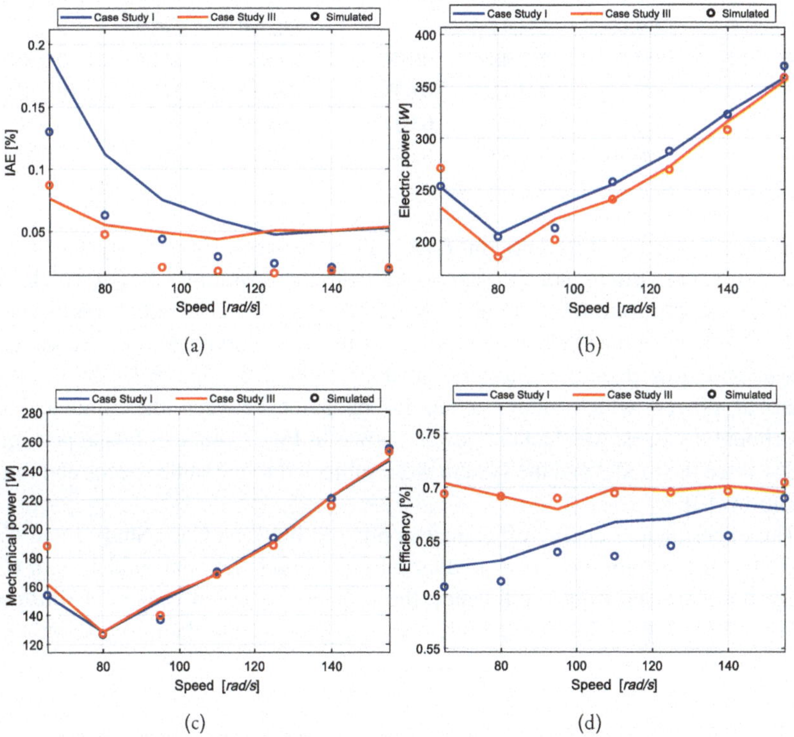

Fig. 7.14 Tests performed for Case Study I and Case Study II with steps applied at 65 rad/s $\leq \omega_{ref} \leq$ 155 rad/s and C_c = [2.31, 1.61, 1.60, 1.53, 1.53, 1.59, 1.6] N · m: (**a**) normalized accumulated error magnitude, (**b**) electrical power, (**c**) mechanical power, and (**d**) energy efficiency

Table 7.9 Accumulated error for tests performed in simulation and on the test bench with Case Study I and Case Study III, within the range 65 rad/s $\leq \omega_{ref} \leq$ 155 rad/s

	Case Study I			Case Study III		
ω_{ref} [rad/s]	IAE_s [%]	IAE_e [%]	Deviation	IAE_s [%]	IAE_e [%]	Deviation
65	13.03	19.26	0.478	8.73	7.67	0.121
80	6.31	11.25	0.782	4.78	5.56	0.163
95	4.39	7.54	0.717	2.15	4.95	1.302
110	2.98	5.95	0.996	1.83	4.40	1.404
125	2.44	4.76	0.950	1.68	5.51	2.279
140	2.12	5.06	1.386	1.88	5.10	1.712
155	1.95	5.29	1.712	2.10	5.39	1.566

Table 7.10 Electrical power for tests performed in simulation and on the test bench with Case Study I and Case Study III, within the range 65 rad/s $\leq \omega_{ref} \leq$ 155 rad/s

	Case Study I			Case Study III		
ω_{ref} [rad/s]	P_{e_s} $[W]$	P_{e_e} $[W]$	Deviation	P_{e_s} $[W]$	P_{e_e} $[W]$	Deviation
65	253.41	252.54	0.003	271.16	233.32	0.139
80	204.72	207.21	0.012	185.73	187.41	0.009
95	213.08	233.41	0.095	202.50	222.28	0.097
110	257.60	255.46	0.008	240.91	241.03	0.0004
125	287.30	285.11	0.007	269.83	272.72	0.010
140	323.24	325.12	0.005	308.50	316.86	0.027
155	370.10	359.83	0.027	359.74	356.77	0.008

In Fig. 7.14a, the difference in IAE values between Case Study I and Case Study III is more pronounced in regions far from the operating point at which Case Study I was optimized (the optimized point being ω_{ref} = 155 rad/s). Furthermore, in Fig. 7.14a, a similar behavior is observed between simulated and experimental values, with low deviations, as presented in Table 7.9. Figure 7.14b shows the electrical power, which shows a similar behavior between the simulated and experimental values. Mechanical power, shown in Fig. 7.14c, exhibits approximate values since the model and the system were subjected to the same speeds and torque conditions [2].

The experimental energy efficiency, compared between Case Study I and Case Study III, follows the behavior predicted by the computational model, with Case Study III achieving higher efficiency throughout the speed range, as illustrated in Fig. 7.14d. Tables 7.9, 7.10, 7.11, and 7.12 present the data corresponding to Fig. 7.14.

Table 7.11 Mechanical power for tests performed in simulation and on the test bench with Case Study I and Case Study III, within the range 65 rad/s $\leq \omega_{ref} \leq$ 155 rad/s

	Case Study I			Case Study III		
ω_{ref} [rad/s]	P_{m_s} [W]	P_{m_e} [W]	Deviation	P_{m_s} [W]	P_{m_e} [W]	Deviation
65	153.91	153.60	0.002	188.26	161.98	0.139
80	127.00	128.81	0.014	127.53	128.86	0.010
95	137.73	149.84	0.087	140.21	152.00	0.084
110	170.25	168.80	0.008	168.21	168.33	0.0007
125	193.42	191.67	0.009	188.48	190.62	0.011
140	220.64	220.12	0.002	215.72	222.35	0.030
155	255.84	247.78	0.031	253.63	248.24	0.021

Table 7.12 Energy efficiency for tests performed in simulation and on the test bench with Case Study I and Case Study III, within the range 65 rad/s $\leq \omega_{ref} \leq$ 155 rad/s

	Case Study I			Case Study III		
ω_{ref} [rad/s]	η_s [%]	η_e [%]	Deviation	η_s [%]	η_e [%]	Deviation
65	60.74	62.53	0.029	69.42	70.42	0.014
80	61.25	63.23	0.032	69.18	69.21	0.0004
95	64.25	65.18	0.014	69.22	68.38	0.012
110	63.60	66.79	0.050	69.49	69.92	0.006
125	64.58	67.10	0.039	69.58	69.76	0.002
140	65.50	68.48	0.045	69.66	70.16	0.007
155	69.13	68.86	0.003	70.47	69.60	0.012

References

1. M.R.C. Reis et al., Optimized techniques for driving and control of the switched reluctance motor to improve efficiency. Control Eng. Practice **90**, 1–18 (2019)
2. W.R.H. Araujo et al., Efficiency enhancement of switched reluctance generator employing optimized control associated with tracking technique. Energies **14**(24), 8388 (2021). https://doi.org/10.3390/en14248388
3. F.A Gomes et al., Heuristic regression method for descriptive data analysis, in *2016 IEEE 16th International Conference on Environmentand Electrical Engineering (EEEIC)* (IEEE. 2016), pp. 1–5
4. L.D.S. Morais et al., Enhancement of the performance of switched reluctance generators in low wind speed conditions using advanced tracking techniques. Int. J. Energy Res. **2024**(1), 5541150 (2024)

Chapter 8
Applications in Control and Generation with Switched Reluctance Machines

Based on these findings, this chapter also presents practical applications developed by the authors throughout their research. A significant portion of the approaches, experiments, and analyses described were originally investigated and discussed in previous works [1–3]. However, not all applications included here have been formally published. For those already presented in scientific articles, this chapter provides a more detailed exposition, expanding discussions and introducing new insights. Consequently, the lower incidence of references to external studies is justified by the authorship of the presented contributions.

8.1 Methodology for Validating an Indirect Torque Detection System

After validating the computational model using parametric regression to represent the inductances of the SRG, considering specific electrical quantities, this section presents the electromagnetic torque values provided by the system shown in Fig. 6.8. The methodology aims to verify whether the mechanical quantities obtained experimentally are consistent with those calculated by the computational model. Since the focus of this study is to optimize the efficiency of the SRG, which depends on mechanical quantities such as torque and speed, as expressed in (6.8) and (6.11), experimental torque measurement becomes necessary. Due to the high cost of commercially available dynamic torque meters, the indirect torque detection system presented in Fig. 6.8 is developed. The tests are carried out with the GCR operating in an open-loop configuration, maintaining constant speed and load. Initially, the SRG is excited with a voltage of 100 V, as shown in the analysis of Fig. 8.1 [2].

As shown in Fig. 8.1, the experimental and simulation values for the torque of the SRG remain consistent under steady-state conditions, demonstrating the effectiveness of the indirect detection system depicted in Fig. 6.8, where the torque

W. P. Calixto et al., *Control and Tracking Techniques for Switched Reluctance Machines*, Power Systems, https://doi.org/10.1007/978-3-031-86727-9_8

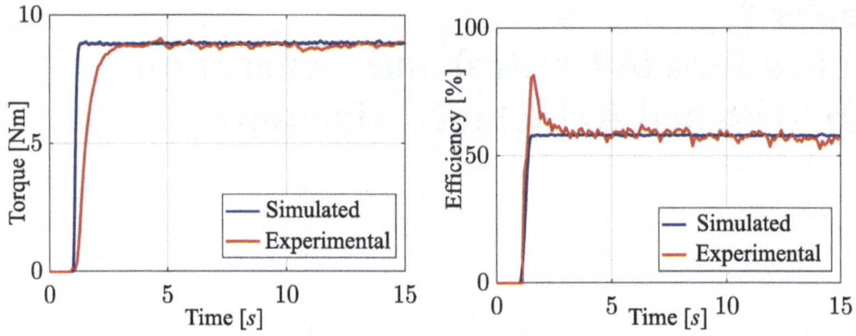

(a) (b)

Fig. 8.1 Evaluation of the mechanical quantities of the computational model compared to experimental data under 100 V excitation: (**a**) torque and (**b**) efficiency

is measured and calculated using (6.7). Regarding the transient behavior of the curves shown in Fig. 8.1a, a divergence is identified in the dynamic response of the generator torque. This divergence occurs because the experimentally implemented indirect detection system calculates the total torque as described in (6.7), while the computational model considers only the friction D and neglects the contribution of the generator's inertia J. According to (2.18), the total torque must also include the term $J \cdot \dfrac{d\omega}{dt}$. However, this term is not taken into account in the computational model, resulting in the observed discrepancy in the transient behavior between the simulated and experimental values.

This exclusion of inertia is maintained because of the absence of an industrially constructed SRG in this study, which complicates the precise operational determination of the machine's moment of inertia. Furthermore, the focus of this study, aimed at optimizing the efficiency of the SRG, requires only the convergence of the experimental and simulation values in steady-state conditions. As illustrated in Fig. 8.1, this convergence is sufficient for both the torque and the efficiency of the machine, as further demonstrated in Fig. 8.1b. Subsequently, the same test is repeated, adjusting the generator's open-loop excitation to 150 V, with the corresponding data presented in Fig. 8.2.

With an increase in generator excitation, the generated power also increases, which subsequently increases torque. In Fig. 8.1a, b, the torque is approximately 8.9 Nm, while the efficiency is around 58%. When the excitation increases to 150 V, the torque and efficiency reach 16 Nm and 52.4%, respectively, as shown in Fig. 8.2a, b. The data derived from the procedures, presented in Figs. 8.1 and 8.2, suggest that the indirect torque detection system described in Fig. 6.8 produces values consistent with the computational model, particularly under steady-state conditions. However, the experimental procedures conducted so far use a fixed conduction window θ_D, with $\theta_{ON} = 0°$ and $\theta_{OFF} = 30°$ [2].

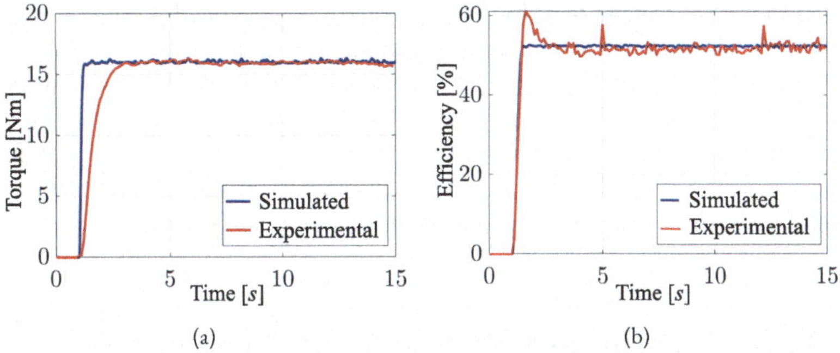

Fig. 8.2 Evaluation of the mechanical quantities of the computational model compared to experimental data under 100 V excitation: (**a**) torque and (**b**) efficiency

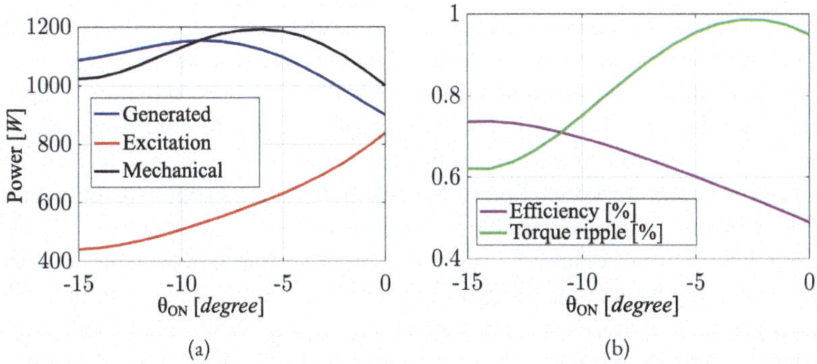

Fig. 8.3 Simulation for the θ_{ON} angle sweep: (**a**) SRG power outputs and (**b**) efficiency and torque ripple

8.2 Conduction Window Simulation in the Power Converter

As indicated in Fig. 5.2, the system variables that can be used as manipulated inputs to control the SRG output voltage include the excitation voltage V_1 and the phase-turn-off angle θ_{OFF}. However, the experimental procedures conducted so far employ a fixed conduction window, as recommended in the literature for 6×4 reluctance machines. Therefore, the study examines whether the advancement of the conduction window improves machine performance, focusing primarily on metrics such as efficiency and torque ripple. Figure 8.3 shows the simulations performed with the θ_{ON} sweep.

Figure 8.3a presents the SRG power curves for θ_{ON} increasing from 0° to $-$15°, while maintaining the conduction window as $\theta_D = \theta_{ON} + 30°$. The power generated corresponds to the electrical output power of the SRG, calculated using (6.10). The excitation power represents the electrical input power to the generator, determined by (6.9), while the mechanical input power is calculated using (6.8). The curves in Fig. 8.3a indicate that, for $\theta_{ON} = -10°$, the generated power exceeds the input powers, ensuring favorable efficiency at this point. In Fig. 8.3b, it is observed that for angles lower than $\theta_{ON} = -10°$, efficiency may increase, but output power decreases. For angles higher than $\theta_{ON} = -10°$, the efficiency decreases and the ripple of the torque increases, creating an unfavorable condition for the generation of energy with the SRG. Therefore, in subsequent methodological steps, the phase turn-on angle θ_{ON} is fixed at $-10°$, as illustrated in Fig. 8.3.

8.3 Analysis of Sensitivity Indices for SRG Parameters

The energy generation by the SRG is adjusted through various variables. Regarding the switching angles, the previous methodology identifies $\theta_{ON} = -10°$ as a favorable point for generation, efficiency, and torque ripple. To evaluate the generator's behavior under variations in electrical and mechanical parameters, sensitivity analysis is applied. The reference values for the variables and their respective variations are shown in Table 6.2. The initial analysis examines the power generated by the SRG in relation to variations in the excitation voltage V_1, turn-off angle θ_{OFF}, and the rotor speed ω. Figure 8.4 illustrates the behavior of the generated power as a function of these variables.

In Fig. 8.4, the horizontal axis represents the baseline power value, defined as 638 W. Analyzing the rate of change of the curves reveals that the excitation voltage V_1 has a greater influence on energy generation compared to the other variables, particularly when V_1 exceeds its baseline value of 150 V. Figure 8.5 illustrates the sensitivity of the SRG efficiency to variations in V_1, θ_{OFF}, and ω.

In Fig. 8.5, the horizontal axis represents the baseline efficiency, corresponding to 0.78. It is observed that a significant increase in the turnoff angle, particularly above the baseline value ($\theta_{OFF} = 15°$), reduces the efficiency of the SRG. On the other hand, variations in excitation voltage have minimal impact on efficiency, except when the variation is below -50%. Figure 8.6 illustrates the sensitivity of the torque ripple to variations in V_1, θ_{OFF}, and ω [2].

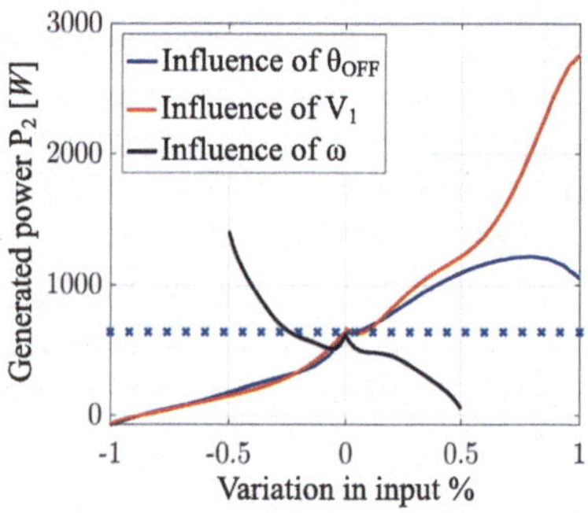

Fig. 8.4 Sensitivity of generated power to variations in V_1, θ_{OFF}, and ω

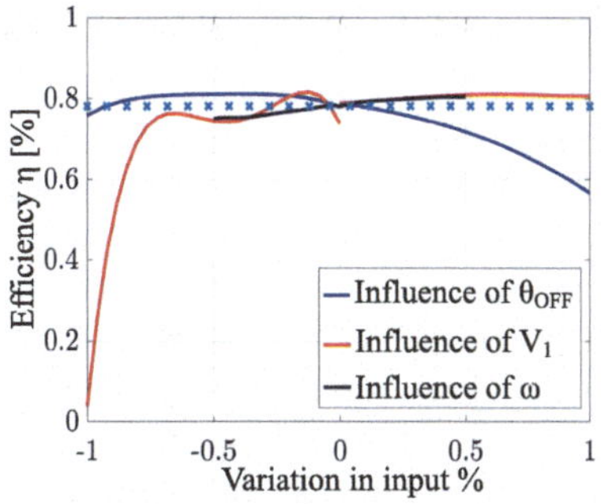

Fig. 8.5 Efficiency sensitivity to variations in V_1, θ_{OFF}, and ω

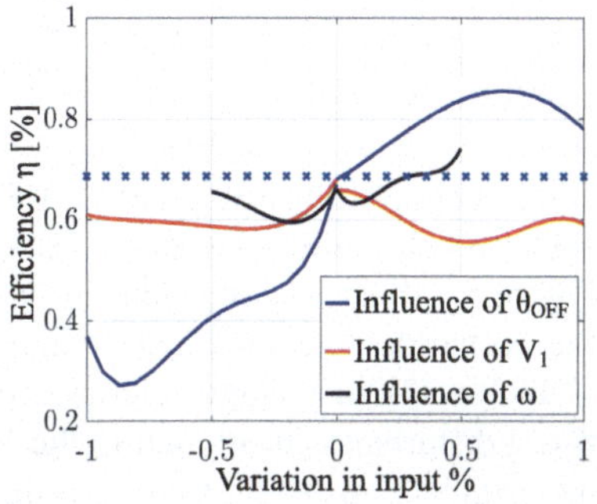

Fig. 8.6 Sensitivity of torque ripple to variations in V_1, θ_{OFF}, and ω

Table 8.1 Sensitivity indices of SRG variables

Analyzed variable	Sensitivity to θ_{OFF}	Sensitivity to V_1	Sensitivity to ω
Generated Power P_2	31.29%	56.51%	12.20%
Efficiency η	31.53%	45.31%	23.16%
Torque Ripple T_R	44.97%	36.94%	18.09%

In Fig. 8.6, the horizontal axis represents the baseline value of T_R, which is 0.69. It is observed that the angle of turn-off significantly influences torque ripple, particularly for values exceeding 50%. This occurs because a larger conduction window, resulting from higher θ_{OFF} values, may overlap with the region where the coil inductance has a rising slope, an area where the machine operates as a motor, leading to increased oscillations in the generator. In addition to the curves presented in Figs. 8.4, 8.5, and 8.6, the sensitivity indices of the analyzed variables are calculated. These indices are determined according to (4.24) and are shown in Table 8.1.

Table 8.1 quantitatively presents the sensitivity of the power, efficiency, and torque ripple to the variables V_1, θ_{OFF}, and ω. The excitation voltage is observed to exert a greater influence compared to the other variables, suggesting that the voltage may be more suitable as a manipulated variable to control the SRG voltage rather than the turn-off angle θ_{OFF}. However, additional experimental procedures are performed to validate this hypothesis.

8.4 Analysis of SRG Voltage PID Controller Optimization

Based on previous methodologies, which indicate that the phase-turn-on angle θ_{ON} contributes positively to the generator when set to $-10°$, and that the excitation voltage V_l shows greater potential to control the generator's output voltage, the optimization process is applied to tune the PID controller. In this context, the adjusted parameters for the PID voltage gain, using the excitation voltage as the manipulated variable, are presented in Table 8.2.

Given the need to evaluate the efficiency of SRG voltage control using either excitation or the turn-off angle as the manipulated variable, the optimization process is applied to both cases. The adjusted parameters for θ_{OFF} as the manipulated variable are presented in Table 8.3.

With the controller gains adjusted for tuning, closed-loop simulations are conducted with a resistive load of 50.6 Ω. The speed is maintained constant at 1000 rpm, and the simulations evaluate the performance of the voltage PID controller for different reference values and disturbances applied to the resistive load. Figure 8.7 illustrates the behavior observed in both simulations and experiments in this scenario, using the SRG excitation voltage as the manipulated variable [2].

In Fig. 8.7, a voltage reference of 80 V is set for the PID controller. At $t = 8\ s$, a disturbance is applied to the load by adding a resistor 94 Ω in parallel with the original resistor 50.6 Ω, resulting in a total resistance of 32.9 Ω. The plots in Fig. 8.7 demonstrates that the optimized gains of the PID controller, as described in Table 8.2, enable the system to maintain the output voltage close to the reference value (80 V) even under load disturbances.

Furthermore, the plots confirm the consistency with the experimental inductance surface shown in Fig. 5.15, as the curves in Fig. 8.7a, b exhibit a similar behavior, indicating a good agreement between the experimental data and the computational model. To evaluate the performance of the PID controller under different conditions, the test is repeated with the reference value adjusted from 80 V to 100 V. The corresponding plots for this test are shown in Fig. 8.8.

As in the previous test, with a reference of 100 V, the PID voltage controller, with gains adjusted through the optimization process, regulates the generator's output voltage according to the established reference and maintains stability under load disturbance. The disturbance, identical to that of the previous test, is applied at $t = 8\ s$. Finally, a closed-loop test is performed using V_l as the manipulated variable,

Table 8.2 Parameters of the voltage PID controller using V_l as the manipulated variable

Proportional gain	Integral gain	Derivative gain
$kp = 84.04 \cdot 10^{-3}\ \circ/V$	$ki = 512.77 \cdot 10^{-3}\ \circ/V$	$kd = 2.146 \cdot 10^{-3}\ \circ/V$

Table 8.3 Parameters of the voltage PID controller using θ_{OFF} as the manipulated variable

Proportional gain	Integral gain	Derivative gain
$kp = 42.86 \cdot 10^{-3}\ \circ/V$	$ki = 297.94 \cdot 10^{-3}\ \circ/V$	$kd = 0.76 \cdot 10^{-3}\ \circ/V$

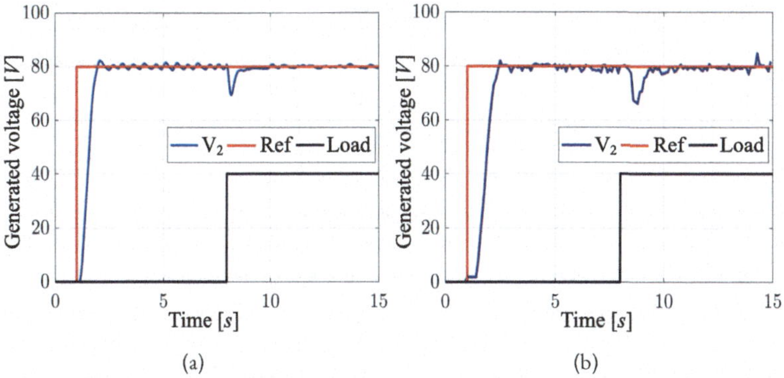

Fig. 8.7 Voltage PID controller using V_2 as the manipulated variable with a reference of 80 V: (**a**) simulation and (**b**) experimental

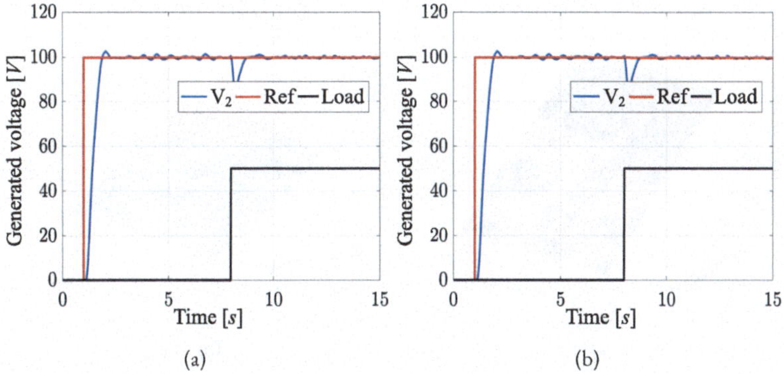

Fig. 8.8 Voltage PID controller using V_2 as the manipulated variable with a reference of 100 V: (**a**) simulation and (**b**) experimental

with the reference adjusted to 120 V. The corresponding plots for this test are shown in Fig. 8.9 [2].

For the 120 V reference, the graphs in Fig. 8.9 demonstrate that the controller maintains consistent performance in both the simulation and the experimental tests. Based on the analyses shown in Figs. 8.7, 8.8, and 8.9, it is decided to use the excitation voltage as a manipulated variable for the output voltage control. Although optimization of the controller using θ_{OFF} as the manipulated variable has already been performed, an additional study is conducted to compare these two alternatives and determine the final choice [2].

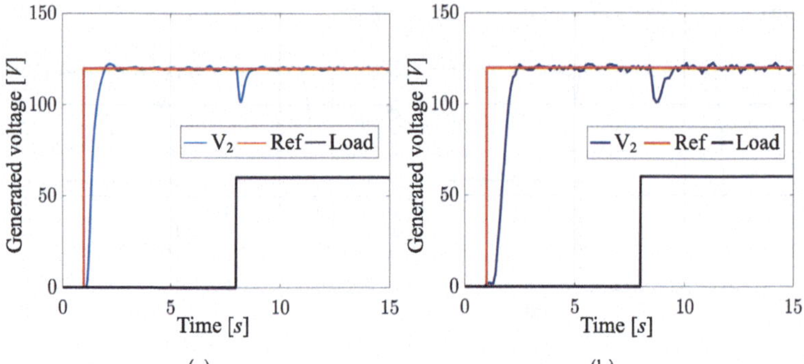

Fig. 8.9 Voltage PID controller using V_2 as the manipulated variable with a reference of 120 V: (**a**) simulation and (**b**) experimental

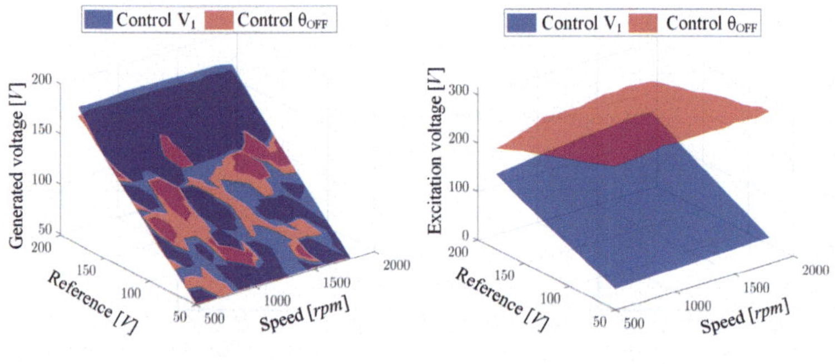

Fig. 8.10 Simulated surfaces for the PID controller acting on V_1 and θ_{OFF}: (**a**) generated voltage V_2 and (**b**) excitation voltage V_1

8.5 Surface Mapping to Identify the Manipulated Variable

The simulation procedures are conducted with the SRG under voltage control, using both the excitation voltage and the turn-off angle as manipulated variables. In both cases, the PID controller is tuned with the gains provided in Tables 8.2 and 8.3. Figure 8.10 illustrates the output voltage of the SRG as well as the excitation voltage applied to the generator by the PID controllers configured for V_1 and θ_{OFF}.

In Fig. 8.10a, the slope of the planes, approximately 45°, indicates that both controllers maintain the generator output voltage close to the reference values. However, the analysis in Fig. 8.10b shows that the PID controller acting on θ_{OFF} applies higher voltages to the DC bus of the power converter. This behavior may be detrimental to the SRG, as the applied voltage can exceed the machine's nominal voltage. This occurs because, with smaller conduction windows, achieving a specific output voltage requires a higher excitation voltage. Figure 8.11 illustrates the

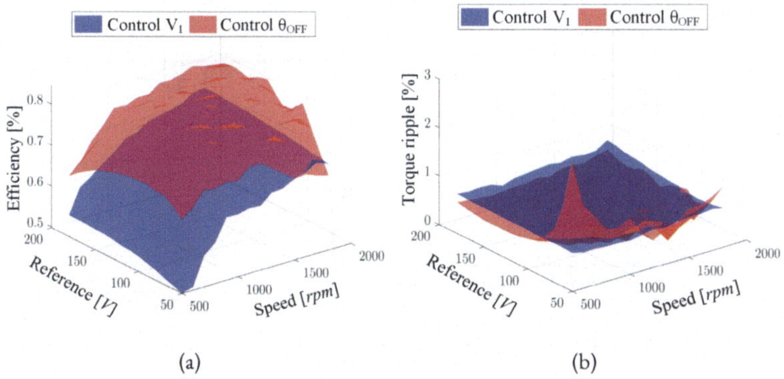

Fig. 8.11 Simulated surfaces for the PID controller acting on V_1 and θ_{OFF}: (**a**) efficiency η and (**b**) torque ripple T_R

efficiency of the SRG and the torque ripple when the generator operates with both PID controllers, considering the action on the excitation voltage and the turn-off angle [3].

In Fig. 8.11a, it is observed that voltage control through the turn-off angle provides higher efficiency for the SRG, while Fig. 8.11b shows that excitation-based control results in greater torque ripple. However, the operational cost associated with the control of the turning angle, as illustrated in Fig. 8.10b, justifies its exclusion, despite offering higher efficiency and lower torque ripple. Consequently, excitation-based control is considered more suitable due to its safety, sensitivity, and, when combined with the tracker, its potential to enhance efficiency equally [3].

8.6 Tracking Technique Performance for Efficiency

The investigation into the possibility of improving the efficiency of the SRG while maintaining control of the output voltage is of great relevance. To avoid modifications to the hardware of the system shown in Fig. 5.2, only the tracker algorithm is developed. Based on previous studies, the angle of turn-on θ_{ON}, the sensitivity of the excitation voltage V_1, and the high levels of excitation applied to the SRG by the control based on θ_{OFF}, this stage involves experimental procedures conducted in both the simulation and the experimental platform shown in Fig. 5.11. Figure 8.12 illustrates the evolution of the SRG switching angles when operated with the MEPT algorithm, as well as the performance of the voltage controller integrated with the MEPT algorithm [3].

In Fig. 8.12a, the switching angles initially remain in the conventional configuration, with $\theta_{ON} = 0°$ and $\theta_{OFF} = 30°$, for a time interval defined. This interval is necessary for the output voltage to reach steady state under the action of the PID controller. Subsequently, the conduction window is adjusted by delaying θ_{ON} in

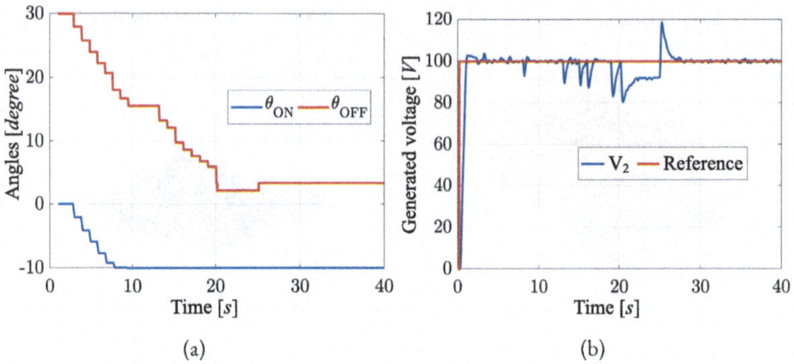

Fig. 8.12 Simulation for the 100 V reference: (**a**) evolution of θ_{ON} and θ_{OFF} angles with the MEPT algorithm and (**b**) performance of the voltage controller in parallel with the MEPT algorithm

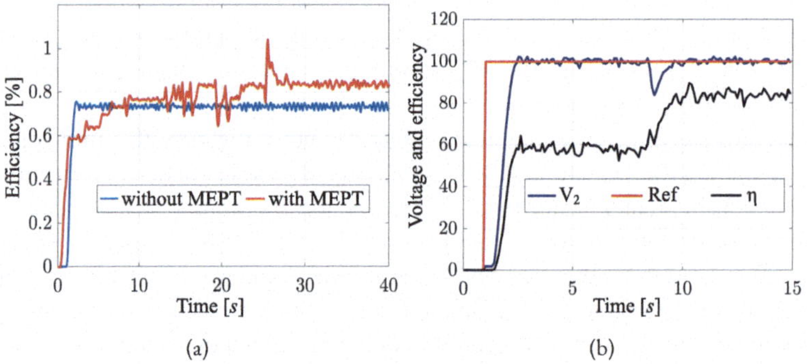

Fig. 8.13 Simulation and experimental analysis for a 100 V reference: (**a**) simulation of efficiency improvement with the MEPT algorithm and (**b**) experimental evaluation of the controller's performance and SRG efficiency

increments of 1° until reaching the limit of $\theta_{ON} = -10°$, previously identified as suitable for the SRG. After this adjustment, the angle of θ_{OFF} is reduced using a perturb-and-observe algorithm, where each degree of reduction is followed by a comparison of the efficiency with the previous value, stopping once no further improvement is achieved. In Fig. 8.12, the angles stabilize at $\theta_{ON} = -10°$ and $\theta_{OFF} = 3°$, resulting in a conduction window of 13°, which reduces the excitation power of the SRG and enhances the efficiency of the generator.

Figure 8.12b illustrates the performance of the voltage PID controller operating in conjunction with the MEPT tracking algorithm, maintaining the stability of the output voltage despite changes in the converter's switching angles. In this test, the reference voltage is set to 100 V. Finally, Fig. 8.13 shows the evolution of the SRG efficiency during the operation of the MEPT algorithm, along with the experimental data obtained with the adjusted angles.

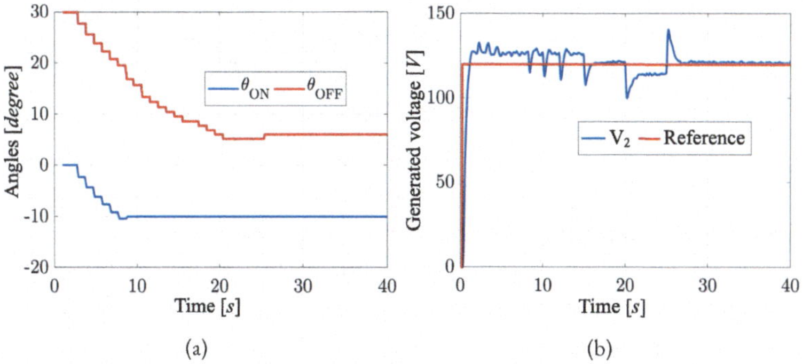

Fig. 8.14 Simulation for a 120 V reference: (**a**) evolution of θ_{ON} and θ_{OFF} angles with the MEPT algorithm, and (**b**) performance of the voltage controller in parallel with the MEPT algorithm

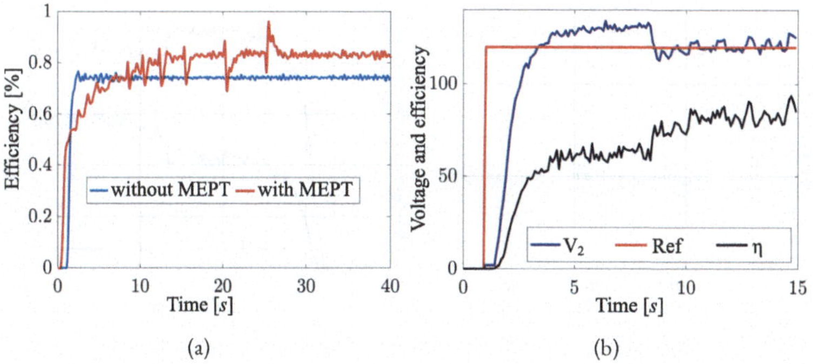

Fig. 8.15 Simulation and experimental analysis for a 120 V reference: (**a**) simulation of efficiency improvement with the MEPT algorithm and (**b**) experimental evaluation of the controller's performance and SRG efficiency

In Fig. 8.13b, the efficiency of the SRG increases from 76% to 83.7%, as determined by the computational model. To experimentally validate the performance of the MEPT algorithm, the angles provided by the algorithm, shown in Fig. 8.12a, are applied to the experimental platform, with the results shown in Fig. 8.13b. In Fig. 8.13b, it is observed that, in addition to the controller maintaining the output voltage at the reference value (100 V), the efficiency reaches approximately 83.7%, as indicated in Fig. 8.13b. Figures 8.14 and 8.15 present the same procedure with a reference voltage of 120 V. In Fig. 8.14a, the switching angles stabilize at $\theta_{ON} = -10°$ and $\theta_{OFF} = 6°$. In Fig. 8.15a, the PID controller maintains the generator output voltage at 120 V, despite transient instability.

Figure 8.15a shows the evolution of the efficiency of the SRG using the MEPT algorithm, with the efficiency increasing from 78.82% to 83.17%. As in the previous procedure, the calculated switching angles, $\theta_{ON} = -10°$ and $\theta_{OFF} = 6°$, are applied to the experimental platform, and the performance of the reference 120 V is shown in Fig. 8.15b. As the final step in the experimental validation of the MEPT

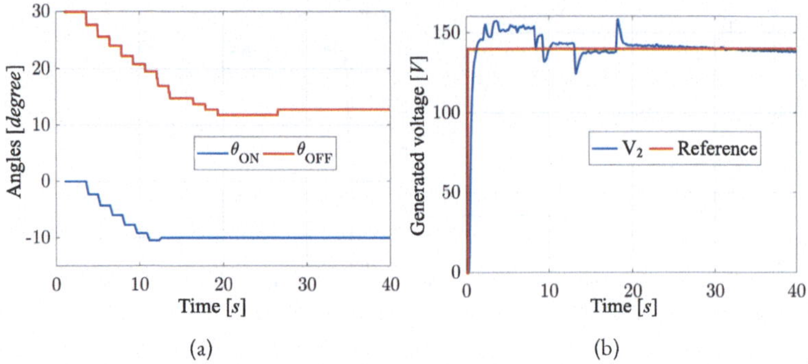

Fig. 8.16 Simulation for a 140 V reference: (**a**) evolution of θ_{ON} and θ_{OFF} angles with the MEPT algorithm and (**b**) performance of the voltage controller in parallel with the MEPT algorithm

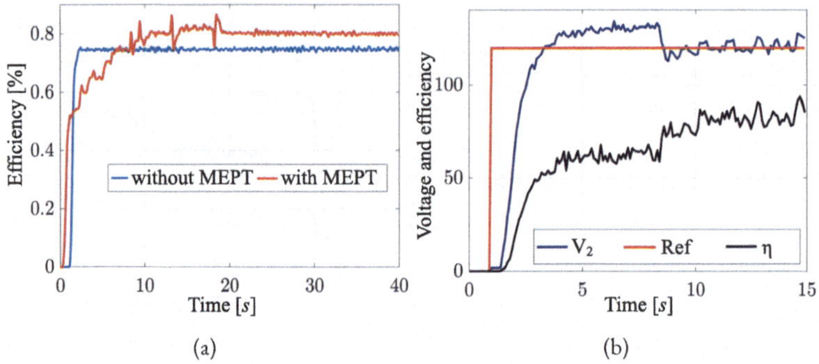

Fig. 8.17 Simulation and experimental analysis for a 140 V reference: (**a**) simulation of efficiency improvement with the MEPT algorithm and (**b**) experimental evaluation of the controller's performance and SRG efficiency

algorithm in conjunction with the PID voltage controller, the reference value is adjusted to 140 V. Figures 8.16 and 8.17 illustrate the behavior observed in this test.

The evolution of the switching angles illustrated in Fig. 8.16a, driven by the MEPT algorithm, stabilizes at $\theta_{ON} = -10°$ and $\theta_{OFF} = 13°$. The performance of the voltage PID controller, shown in Fig. 8.16b, demonstrates consistent operation in conjunction with the MEPT algorithm, considering the adjustments made through the optimization process. The generator output remains close to the reference value of 140 V. In Fig. 8.17a, the efficiency of the SRG increases from 77% to 80.3%, as indicated by the simulation data. Finally, the angles $\theta_{ON} = -10°$ and $\theta_{OFF} = 13°$, determined by the tracking algorithm, are applied to the experimental platform, and the corresponding performance is shown in Fig. 8.17b, highlighting an improvement in generator efficiency [3].

8.7 Wind Energy Generation with the Switched Reluctance Generator

8.7.1 *A Tracker Designed for θ_{off} in the Independently Excited Model*

The value $\theta_{on} = -10°$ is selected as a viable option to ensure efficient generation, high efficiency, and reduced torque ripple. Keeping $V_e = 100\,\text{V}$ and with the aim of optimizing the system, simulations are carried out for $100\,\text{rpm} \leq \omega_g \leq 2000\,\text{rpm}$ and $-5° \leq \theta_{off} \leq 30°$, as illustrated in Fig. 6.20. Based on the power and efficiency curves, the value of θ_{off} is identified that maximizes the output power P_s. However, since the SRG requires an external excitation source, the input power P_e, which represents the energy supplied to the excitation source, is also evaluated. Therefore, the proposed optimization does not aim to achieve maximum power or track the maximum power point, as MPPT algorithms do. Instead, the objective is to determine the value θ_{off} for each rotational speed ω_g of the SRG that achieves both the maximum generated power and the maximum achievable efficiency.

Figure 8.18 shows that the input electrical power P_e is nearly zero for the lowest θ_{off} values, particularly in the range of -5° to 10°. This behavior occurs because θ_{on} is fixed at -10°, and reduced θ_{off} values lead to a narrow conduction window θ_D. Consequently, each SRG coil is energized for a short period. Furthermore, P_e remains below 2000 W at all rotational speeds when $\theta_{off} < 13°$. Beyond this value, P_e reaches its highest levels for $\omega_g < 500\,\text{rpm}$, peaking at $\omega_g = 100\,\text{rpm}$ and $\theta_{off} = 30°$.

Similarly to P_e, P_{mec}, shown in Fig. 8.19, is also affected when the values of θ_{off} are below 15°. An increase in P_{mec} is observed as θ_{off} grows; however, this increase occurs only up to $\omega_g \approx 1200\,\text{rpm}$, after which P_{mec} begins to decrease. Regarding P_s, illustrated in Fig. 8.20, an increase in θ_{off} leads to higher values of P_s. However, the optimal values of θ_{off} vary with ω_g, indicating that achieving the maximum generation efficiency requires dynamic values θ_{off}. In addition to P_{mec}, P_e, and P_s, the efficiency η, shown in Fig. 8.21, is also analyzed. It is observed that η is higher for greater values of ω_g. Thus, the optimization process aims to

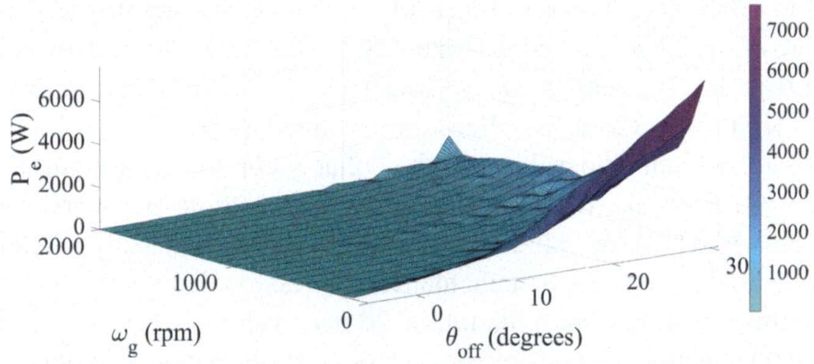

Fig. 8.18 $P_e \times \theta_{off} \times \omega_g$

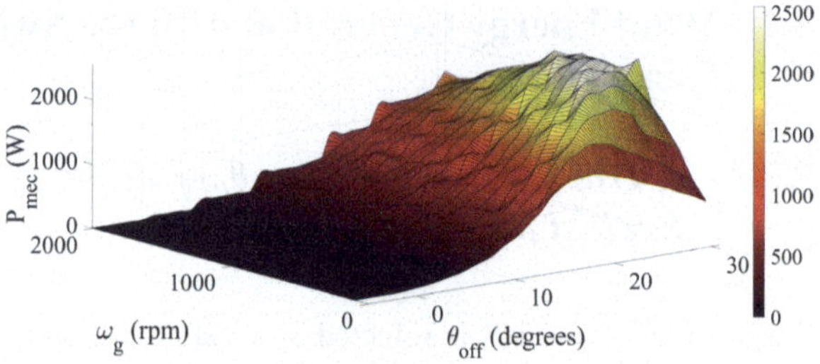

Fig. 8.19 $P_{mec} \times \theta_{off} \times \omega_g$

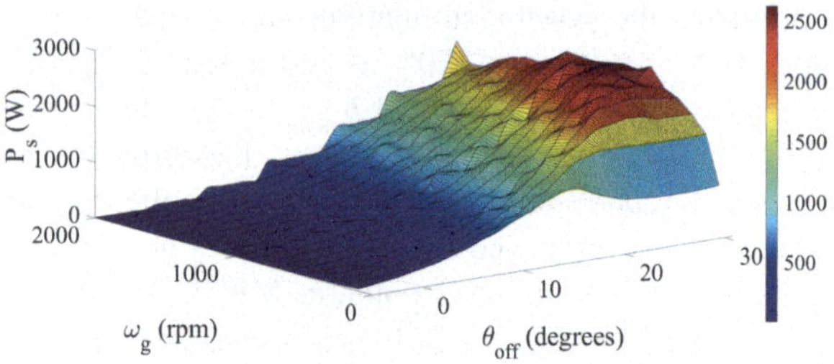

Fig. 8.20 $P_s \times \theta_{off} \times \omega_g$

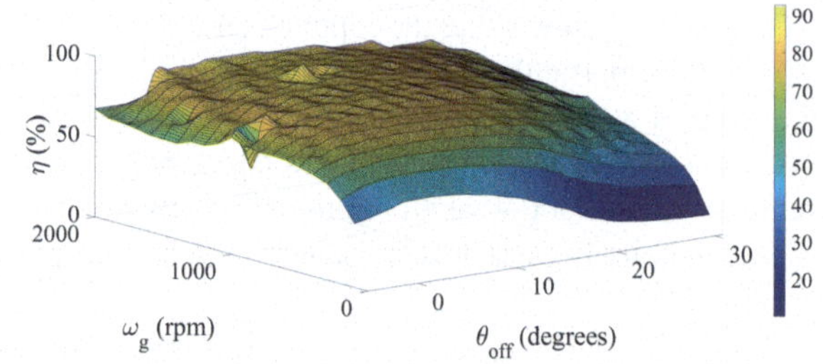

Fig. 8.21 $\eta \times \theta_{off} \times \omega_g$

determine the θ_{off} angles that produce the maximum P_s and the highest possible η for each generator operating speed.

This approach is adopted considering that, when seeking θ_{off} values that maximize P_s, P_e may, in some cases, reach high levels, leading to $\eta \approx 0$, which makes the operation impossible. In contrast, when considering only the value θ_{off} that maximizes η, P_s can, in some cases, approach 0, which also renders the operation unfeasible. The analysis of power surfaces, shown in Figs. 8.19 and 8.20, in relation to efficiency, shown in Fig. 8.21, allows identifying the θ_{off} angles that simultaneously optimize P_s and η. Figure 8.22 illustrates the relationship among P_s, η, and θ_{off} for $100\,\text{rpm} \leq \omega_g \leq 2000\,\text{rpm}$. To facilitate interpretation, P_s is normalized with respect to its baseline value, defined as 2626 W.

Taking into account the values of θ_{off} that yield the maximum P_s and the maximum η for each ω_g, these points can be identified at the intersection of the respective curves. Thus, the value θ_{off} is determined for each analyzed range of ω_g. However, the goal is to derive a mathematical expression that relates any ω_g to θ_{off}. To achieve this, an initial interpolation of 20 θ_{off} values is performed, producing 400 interpolated values. Based on these data, a ninth degree polynomial is fitted using the least squares method. The polynomial $\theta_{off}(\omega_g)$ is expressed as:

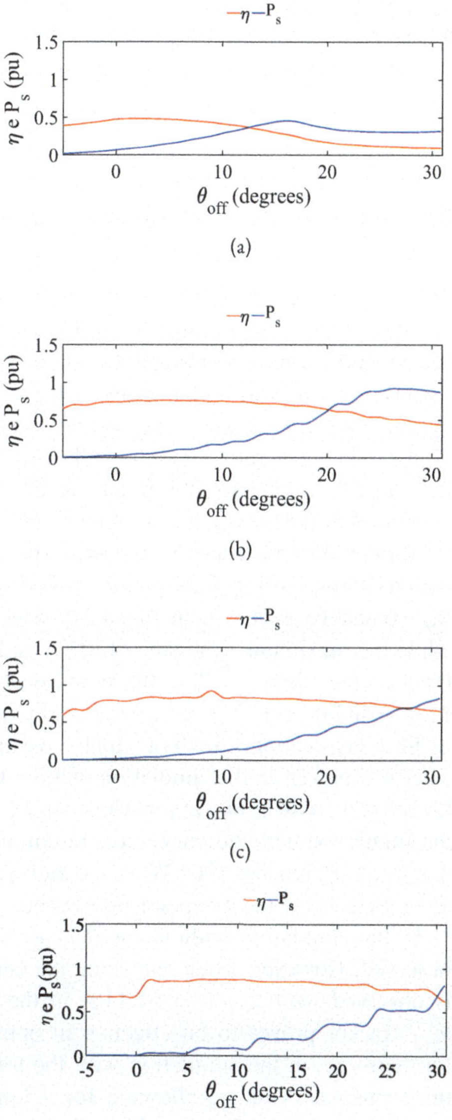

Fig. 8.22 Electrical power output and system efficiency for different values of ω_g: (**a**) $\omega_g = 100$ rpm, (**b**) $\omega_g = 500$ rpm, (**c**) $\omega_g = 1500$ rpm, and (**d**) $\omega_g = 2000$ rpm

$$\begin{aligned}\theta_{off}(\omega_g) = &+ 1.031 \times 10^{-26}\omega_g^9 - 9.765 \times 10^{-23}\omega_g^8 + 3.920 \times 10^{-19}\omega_g^7 \\ &- 8.757 \times 10^{-16}\omega_g^6 + 1.207 \times 10^{-12}\omega_g^5 - 1.087 \times 10^{-9}\omega_g^4 \\ &+ 6.693 \times 10^{-7}\omega_g^3 - 0.000286\omega_g^2 + 0.0872\omega_g + 5.973\end{aligned} \tag{8.1}$$

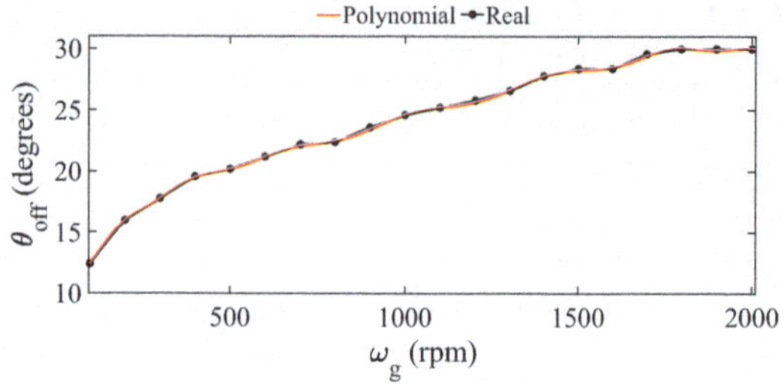

Fig. 8.23 Polynomial × real data for the SRGIE

From (8.1), the optimal θ_{off} values covering the entire range of ω_g are determined, as shown in Fig. 8.23. These values are associated with different wind speeds (v_v) from various locations. Thus, it is possible to identify the values θ_{off} that optimize system operation. The expression (8.1) is then integrated in the simulation of the wind generation system, allowing the determination of the ideal values θ_{off} and, consequently, the maintenance of the optimal ratio λ. This ensures the extraction of the maximum mechanical power from the wind turbine.

Simulations are carried out using the v_v profiles corresponding to the three regions shown in Fig. 6.28 and described in Table 6.6. In simulation **without the θ_{off} tracker**, as shown in Fig. 8.24b, the minimum P_e is 543 W in Location 1, while the maximum P_e reaches 1303 W in Location 2. In simulation **with the θ_{off} tracker**, the minimum P_e is 768 W in Location 1, whereas the maximum P_e reaches 1134 W in Location 2. The θ_{off} tracker increases P_e by 35.27% in Location 3 and by 21.96% in Location 1, both of which have higher wind speeds. Regarding the output electrical power, in the simulation without the tracker, $P_s = 1268$ W is obtained in Location 1, with a maximum value of $P_s = 1724$ W recorded in Location 2. In the simulation with the tracker, the minimum P_s is 1551 W in Location 2, while the maximum P_s reaches 1701 W in Location 1. Figure 8.24 illustrates the power output and efficiency of the independently excited SRG with θ_{off} tracking [3].

In the simulation without the θ_{off} tracker, the maximum peak value of P_s is observed. However, when analyzing the entire simulation, a reduction in variation is observed, with $\Delta_{P_s} = 67.11\%$ in the simulation with the θ_{off} tracker. The θ_{off} tracker proves to be effective in optimizing generation (energy utilization). Furthermore, in the simulation with the tracker, the algorithm adjusts θ_D according to increases in v_v, allowing for a longer excitation period of the machine. Mechanical power P_{mec} is also influenced by v_v, as ω_g is adjusted by the θ_{off} tracker. Consequently, in the tracker simulation, the P_{mec} values are higher in regions with a greater v_v. Regarding efficiency η, the recorded values range from $59\% \leq \eta \leq 75.6\%$ in the simulation without the tracker to $61\% \leq \eta \leq 72\%$ in the simulation with the tracker, considering minimum and maximum values. Although the increase in η is not significant, there is a reduction in variation, from $\Delta_\eta = 16.6\%$ without tracking to $\Delta_\eta = 11\%$ with tracking.

Tables 8.4 and 8.5 present the average values of P_e, P_{mec}, P_s, and efficiency η under conditions without and with tracking. An increase in P_s is observed in the three locations when θ_{off} tracking is applied. Although the tracking process aims to

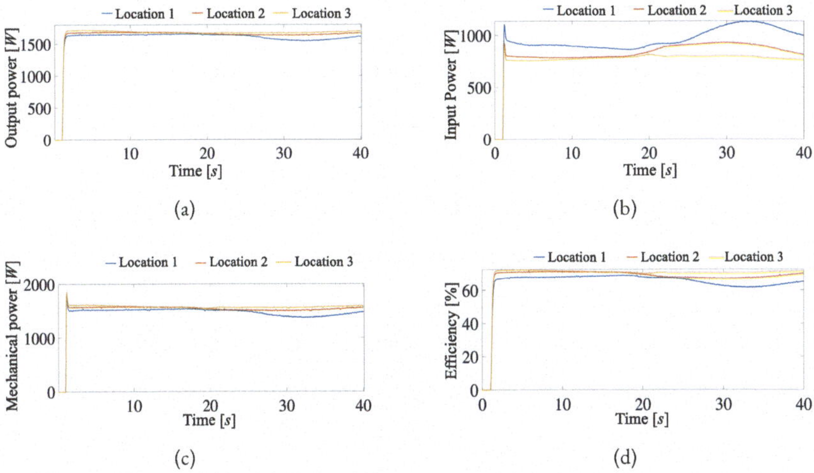

Fig. 8.24 Power outputs and efficiency of the independently excited SRG with θ_{off} tracking: **(a)** P_e, **(b)** P_s, **(c)** P_{mec}, and **(d)** η

Table 8.4 Average without tracking and independent excitation

	P_e	P_{mec}	P_s	η
Location 1	903W	1387W	1522W	**66.13%**
Location 2	679W	1241W	1376W	**71.05%**
Location 3	574W	1155W	1286W	**73.67%**

Table 8.5 Average with tracking and independent excitation

	P_e	P_{mec}	P_s	η
Location 1	941W	1472W	**1600W**	65.50%
Location 2	829W	1533W	**1640W**	68.68%
Location 3	776W	1566W	**1655W**	70.35%

identify θ_{off} values that maximize P_s and η, the simulation without tracking shows a higher global η. However, even with tracking, the values of η remain above 65%, and the values of θ_{off} selected during tracking lead to a significant increase in P_s for the three locations analyzed.

To implement the control of the output voltage of the SRG V_s alongside the θ_{off} tracker, a proportional-integral-derivative (PID) controller is incorporated into the simulation. The value of θ_{on} is kept constant at -10°, and the excitation voltage V_e is selected as the controllable variable. The Nelder–Mead Simplex algorithm is used to optimize the PID controller gains. The reference voltage is set to $V_{s_{ref}} = 180\,\text{V}$, and the initial parameters, based on previous knowledge of the system, are $k_p = 3°/\text{V}$, $k_i = 5°/\text{V}$, and $k_d = 0.1°/\text{V}$. The optimization process determines the gains $k_p = 2.3889°/\text{V}$, $k_i = 6.6075°/\text{V}$, and $k_d = 0.0752°/\text{V}$, with an error deviation of $IAE = 20.2185\,\text{V}$. Figure 8.25 illustrates the simulation for $v_v = [7,\ 10,\ 6] m/s$ at intervals of $3s$.

Fig. 8.25 Output voltage control with variable speed: (**a**) V_s and V_{ref}, and (**b**) ω_g

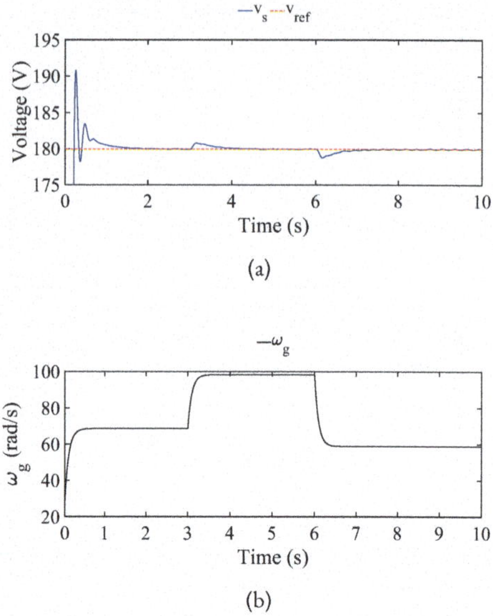

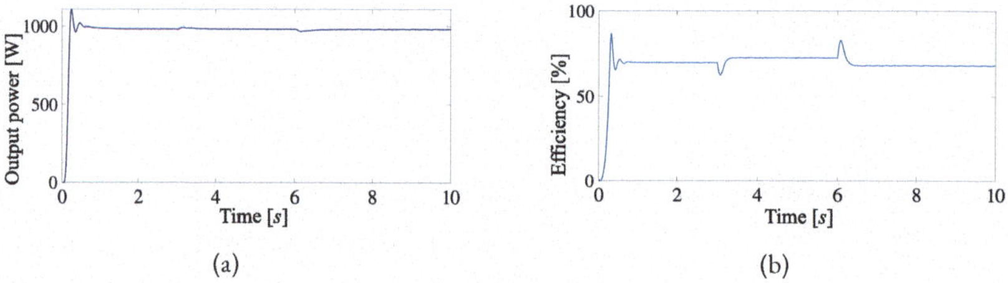

Fig. 8.26 Output voltage control with variable speed: (**a**) V_s and V_{ref} and (**b**) ω_g

The control of V_s is performed through V_e, and it is observed that throughout the simulation, V_s remains close to $V_{s_{ref}} = 180\,\text{V}$. At $t = 3s$, the generator increases ω_g to match the change in v_v, which transitions from $7m/s$ to $10m/s$, and a similar adjustment occurs at $t = 6s$. Despite the changes in ω_g, the PID controller maintains V_s nearly constant. Figure 8.26 illustrates P_s and η under V_s control. Even with variations in ω_g, P_s remains around 1000 W, and η does not show significant changes. The recorded minimum value is $\eta = 68.14\%$, while the maximum is $\eta = 72.89\%$. Following the variable-speed simulations, an additional simulation is conducted with changes in the reference value (setpoint). Initially, $V_{s_{ref}} = 180\,\text{V}$, followed by a change to $V_{s_{ref}} = 220\,\text{V}$ at $t = 3s$, and another adjustment to $V_{s_{ref}} = 160\,\text{V}$ at $t = 6s$. Figure 8.27 illustrates V_s and $V_{s_{ref}}$.

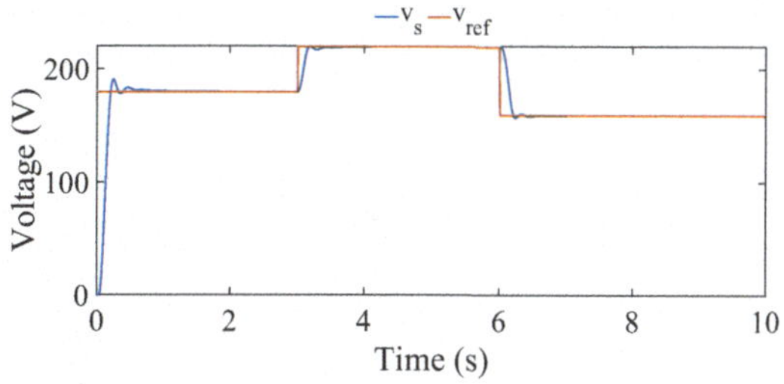

Fig. 8.27 V_s and $V_{s_{ref}}$ under V_s control

8.7.2 Self-excited Switched Reluctance Generator

In the simulation with variations in $V_{s_{ref}}$ and ω_g, the PID controller with optimized gains demonstrates the ability to maintain V_s constant. Furthermore, the PID controller operates in conjunction with the θ_{off} tracker. For the self-excited switched reluctance generator (SRGA) simulation, the same independently excited model is used, with the difference that the SRGA model includes a capacitor responsible for the machine's initial excitation. Furthermore, there is no separation between the excitation and generation circuits, as illustrated in Fig. 2.13. The initial capacitor charge is performed by an external excitation source, which can be a battery or solar energy [3].

After the capacitor's initial charging, the AHB converter switching process and the excitation stage of the coils begin. When the voltage at the coil terminals exceeds the initial capacitor voltage, the diode D_7 becomes reverse biased, and the generator operates entirely in self-excited mode. To prevent interference from the excitation source during the rest of the simulation, an open/closed switch is introduced, which opens after $0.5s$. The capacitor value is set to $5mF$, initially charged to 50 V. The speed $\omega_g = 600$ rpm is kept constant, and the conduction window is fixed at $0° \leq \theta_D \leq 30°$.

Initially, a simulation is performed using the same value of $R_{carga} = 33\Omega$ as in the independently excited model. Figure 8.28 presents the output voltage obtained in this simulation. Analyzing Fig. 8.28, it is observed that the SRGA model is unable to generate energy with $R_{carga} = 33\Omega$ because V_s remains below V_e, causing no energy generation. To evaluate the performance of the SRGA with different values of R_{carga}, simulations are performed considering the range $10\Omega \leq R_{carga} \leq 120\Omega$, in increments of 10Ω, and the range $100\,\text{rpm} \leq \omega_g \leq 2000\,\text{rpm}$, in increments of 100 rpm. The initial capacitor charging voltage is maintained at 50 V to ensure excitation throughout the ω_g range. Using data from these simulations, the P_s values are calculated for all R_{carga} and ω_g values. Figure 8.29 presents the surface generated from the values P_{mec}, R_{carga}, and ω_g.

Figure 8.29 shows that the output electrical power (P_s) increases as the values of R_{carga} and ω_g increase. For R_{carga} values below 60Ω, the machine reaches a maximum of $P_s = 1000$ W throughout the range of ω_g analyzed. With $60\Omega \leq R_{carga} \leq 120\Omega$, a significant increase in P_s is observed, which also increases with

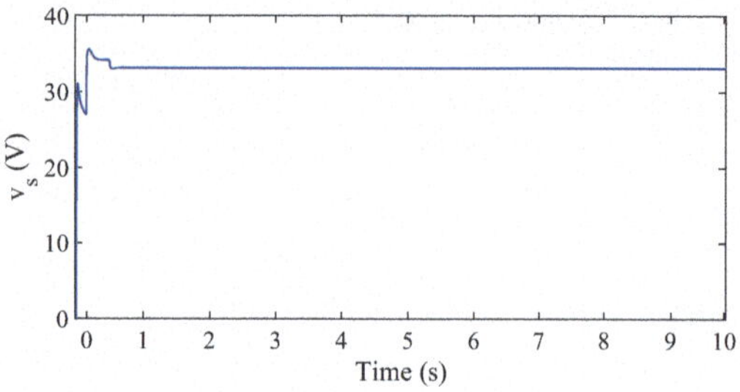

Fig. 8.28 Output voltage of the self-excited generator

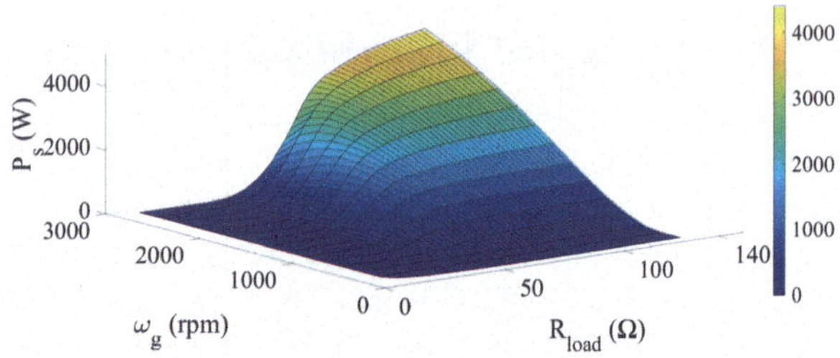

Fig. 8.29 $P_s \times R_{carga} \times \omega_g$

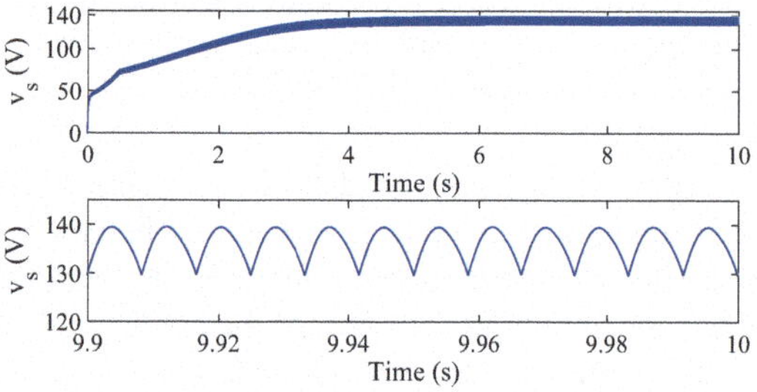

Fig. 8.30 Behavior of the output voltage of the self-excited generator

increasing ω_g, reaching values above $P_s = 4000\,\text{W}$. The actual machine, whose data are used in these simulations, has a power limitation of 2.6 kW. Therefore, $R_{carga} = 100\Omega$ is established, enabling $P_s \approx 2.6\,\text{kW}$ at $\omega_g \approx 1500\,\text{rpm}$. Figure 8.30 presents the output voltage from this simulation. Starting with $V_e = 50\,\text{V}$, the machine operates in self-excitation mode, reaching voltage peaks of approximately $V_s \approx 140\,\text{V}$. In addition, the moments when the capacitor charges and discharges are observed [3].

The inductance curves of the machine's three phases are analyzed using the same method applied to the independently excited model. Figure 8.31 shows the behavior of the voltage, current, command pulse, and inductance for one phase of the SRGA. During the excitation stage of the phases, $V_e \approx 138\,\text{V}$ is observed, and after the switches are turned off, in the generation stage, $V_e \approx 232\,\text{V}$, the current reaching a peak of approximately $25A$. Figure 8.32 presents the performance of P_{mec}, P_e, and P_s for SRGA. After the initial excitation of the capacitor, P_e remains zero. The SRGA achieves $P_{mec} \approx 1200\,\text{W}$ and $P_s \approx 184\,\text{W}$.

Figure 8.33 shows the efficiency η, calculated considering P_{mec} and P_s, as there is no external excitation. With $\eta = 18\%$, the efficiency is relatively low compared to

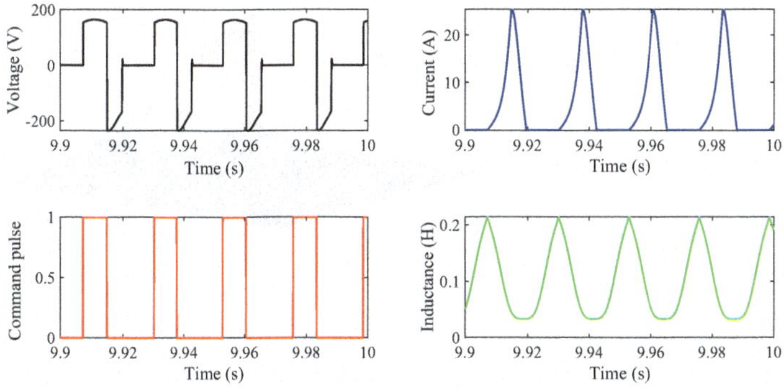

Fig. 8.31 Behavior of one phase of the SRGA

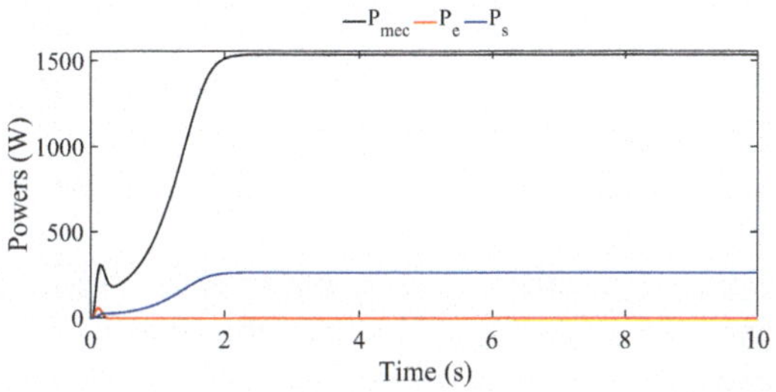

Fig. 8.32 Mechanical power, input electrical power, and output electrical power of the SRGA

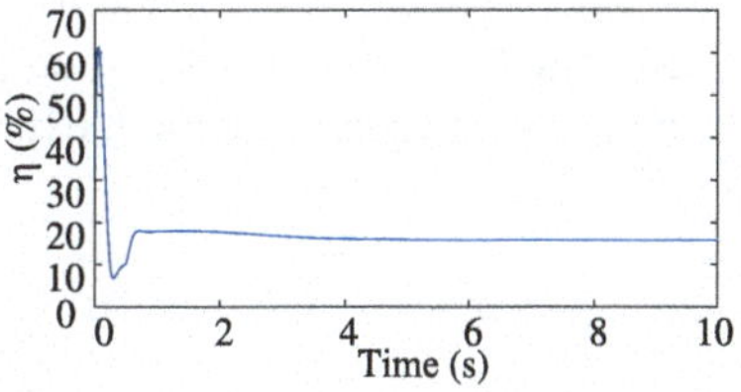

Fig. 8.33 Efficiency of the SRGA

the independently excited model, which achieves $\eta > 50\%$. However, it is important to note that this configuration does not use an external excitation source, providing a significant advantage for wind generation applications.

The θ_{off} tracking for the self-excited model is conducted similarly to the independently excited model. The parameters are set as $\theta_{on} = -10°$, $V_e = 50\,\text{V}$ for the initial charge of the capacitor, and $R_{carga} = 100\Omega$. To optimize the system, simulations are performed considering $100\,\text{rpm} \leq \omega_g \leq 2000\,\text{rpm}$ and $-5° \leq \theta_{off} \leq 30°$. Figure 8.34 shows the surface that relates $P_s \times \omega_g \times \theta_{off}$. In this simulation, P_s reaches values that exceed the nominal electrical power of the machine. Thus, the maximum P_s is defined as 2626 W, highlighted in the red plane in Fig. 8.35.

The determination of θ_{off} for the self-excited model differs from the independently excited model. Since the self-excited model does not rely on an external excitation source, the θ_{off} value selected for each ω_g is the one that achieves the

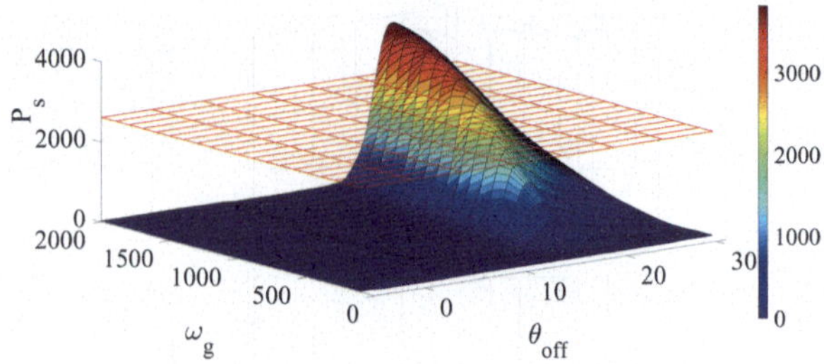

Fig. 8.34 Relationship between $P_s \times \omega_g \times \theta_{off}$

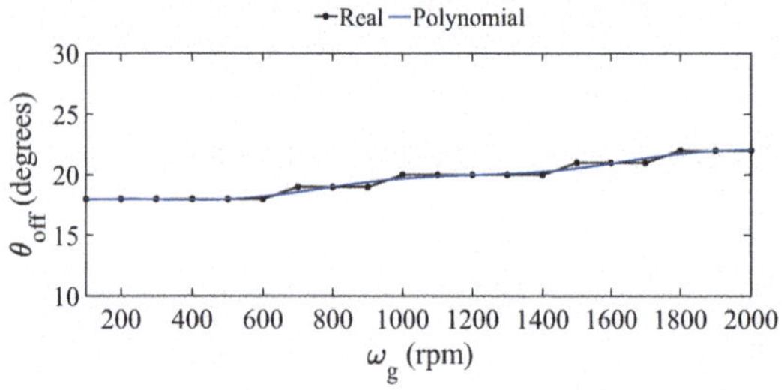

Fig. 8.35 $\theta_{offAE}(\omega_g) \times$ real data for the self-excited model

highest P_s, without considering η. Thus, a specific value of θ_{off} is defined for each analyzed range of ω_g. However, the objective is to derive an expression that relates each ω_g to its corresponding θ_{off}. Before obtaining the polynomial expression, the 20 values θ_{off} are interpolated to generate 400 interpolated values. Based on this interpolated curve, a ninth degree polynomial is fitted for the self-excited model using the least squares method, expressed as:

$$\begin{aligned}\theta_{offAE}(\omega_g) = &- 6.648 \times 10^{-27}\omega_g^{10} + 7.190 \times 10^{-23}\omega_g^{9} - 3.221 \times 10^{-19}\omega_g^{8}\\ &+ 7.706 \times 10^{-16}\omega_g^{7} - 1.060 \times 10^{-12}\omega_g^{6} + 8.397 \times 10^{-10}\omega_g^{5}\\ &- 3.660 \times 10^{-7}\omega_g^{4} + 8.219 \times 10^{-5}\omega_g^{3} - 0.008\omega_g^{2} + 18.307\end{aligned} \tag{8.2}$$

From expression (8.2), the optimal θ_{off} values for the entire range of ω_g operated by SRGA are determined. Figure 8.35 shows the relationship between the actual values of θ_{off} and those determined by $\theta_{offAE}(\omega_g)$ for the v_v profile of Location 1, as described in Table 6.6. Thus, the expression (8.2) $\theta_{offAE}(\omega_g)$ is incorporated into the simulation of the wind generation system, allowing the optimal θ_{off} to be tracked in the same way as for the independent excited model. The θ_{off} tracking is performed for each v_v profile shown in Fig. 6.28. Figure 8.36 presents P_s corresponding to these profiles, with and without θ_{off} tracking. The hypothetical generation is exemplified for: (i) Location 1, (ii) Location 3, and (iii) Location 1. For comparison, Fig. 8.37 shows P_s associated with the same profiles, also with and without θ_{off} tracking, but for the independently excited SRG.

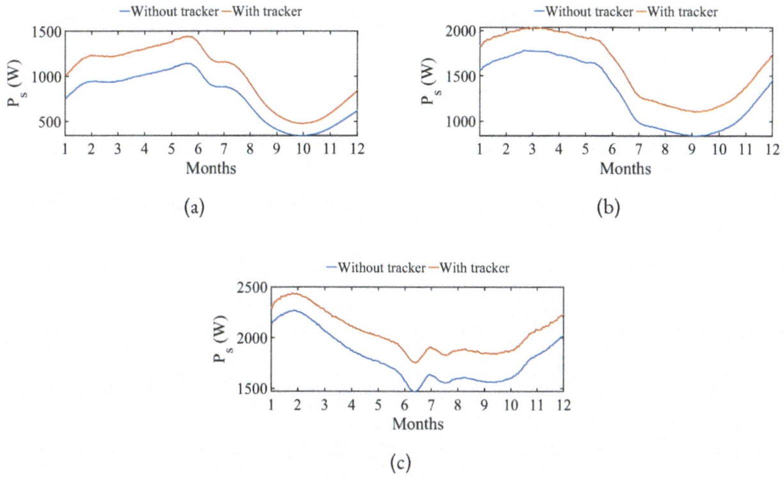

Fig. 8.36 P_s for the three v_v profiles with and without θ_{off} tracking for the self-excited model: (**a**) Location 1, (**b**) Location 2, and (**c**) Location 3

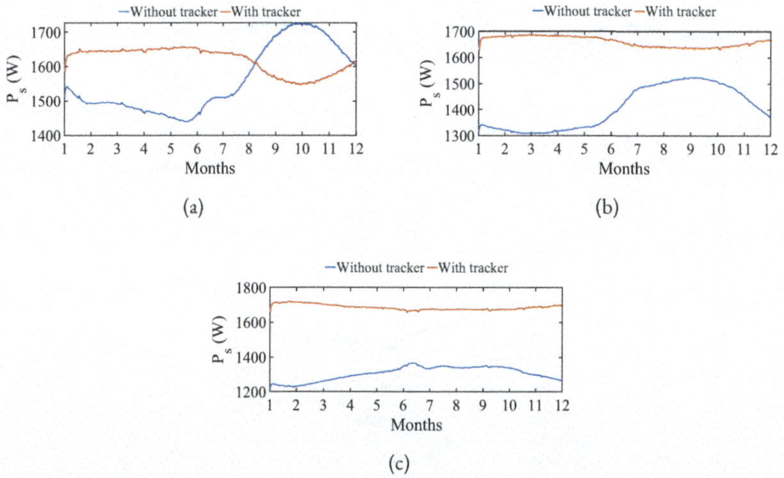

Fig. 8.37 P_s for the three v_v profiles with and without θ_{off} tracking for the independently excited model: (**a**) Location 1, (**b**) Location 2, and (**c**) Location 3

P_s with θ_{off} tracking shows improved performance in both cases, for independent excited and self-excited models, highlighting the effectiveness of the tracker combined with optimized parameters. Figures 8.38 and 8.39 present η for the three v_v profiles in the models independently excited and self-excited. It is observed that,

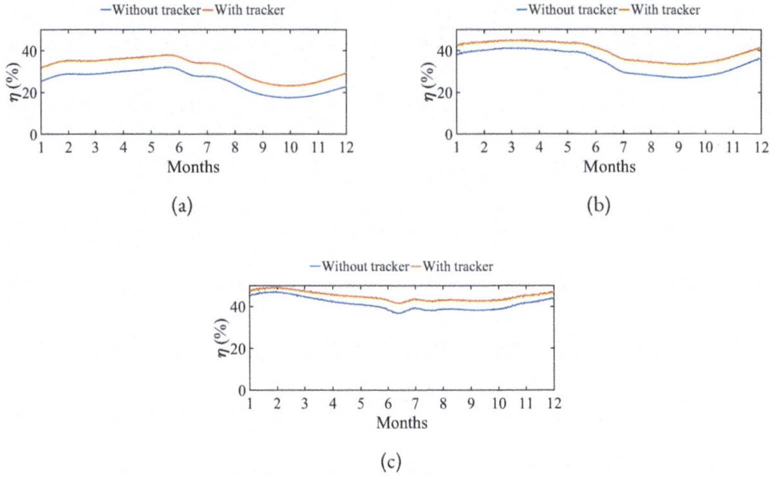

Fig. 8.38 η for the three v_v profiles with and without θ_{off} tracking for the self-excited model: (**a**) Location 1, (**b**) Location 2, and (**c**) Location 3

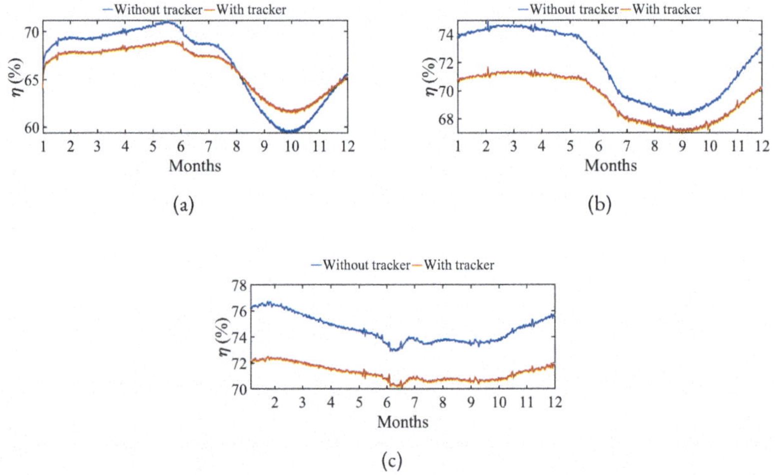

Fig. 8.39 η for the three v_v profiles with and without θ_{off} tracking for the independently excited model: (**a**) Location 1, (**b**) Location 2, and (**c**) Location 3

although the main objective of θ_{off} tracking in the self-excited model is to maximize P_s, the tracker also contributes to an increase in η. In contrast, in the independently excited model, while power maximization is achieved, no significant improvement in η is observed.

References

1. M.R.C. Reis et al., Optimized techniques for driving and control of the switched reluctance motor to improve efficiency. Control Eng. Practice **90**, 1–18 (2019)
2. W.R.H. Araujo et al., Efficiency enhancement of switched reluctance generator employing optimized control associated with tracking technique. Energies **14**(24), 8388 (2021). https://doi.org/10.3390/en14248388
3. L.D.S. Morais et al., Enhancement of the performance of switched reluctance generators in low wind speed conditions using advanced tracking techniques. Int. J. Energy Res. **2024**(1), 5541150 (2024)

Index

W. P. Calixto et al., *Control and Tracking Techniques for Switched Reluctance Machines*, Power Systems, https://doi.org/10.1007/978-3-031-86727-9

The manufacturer's authorised representative in the EU is Springer Nature Customer Service Centre GmbH, Europaplatz 3, 69115 Heidelberg, Germany. If you have any concerns regarding our products, please contact ProductSafety@springernature.com

Printed and bound by CPI Group (UK) Ltd, Croydon, CR0 4YY
30/09/2026
02225592-0002